THE LIBRARY
ST. MARY'S COLLEGE OF MARYLAND
ST. MARY'S CITY, MARYLAND 20686

Environmental Impact Analysis

Peter E. Black

Library of Congress Cataloging in Publication Data

Black, Peter E.
  Environmental impact analysis.

  Bibliography.
  Includes index.
  1. Environmental impact analysis—United States.
I. Title.
TD194.6.B53   333.7'1'0973   81-7339
ISBN 0-03-059618-1   AACR2

Quotation from *Silent Spring* by Rachel
Carson copyright © 1962 by Rachel Carson.
Used with permission of Houghton Mifflin Company.

Published in 1981 by Praeger Publishers
CBS Educational and Professional Publishing
a Division of CBS Inc.
521 Fifth Avenue, New York, New York 10175 U.S.A.

© 1981 by Praeger Publishers

All rights reserved

23456789 052 98765432
Printed in the United States of America

to my parents,
Elinor Goldmark Black and Algernon D. Black,
and to all those who similarly embrace
the words of Thomas Jefferson:
I know of no safe depository of the ultimate powers
of society but the people themselves; and if we think them
not enlightened enough to exercise their control with a
wholesome discretion, the remedy is not to take it
from them, but to inform their discretion.

Contents

List of Tables and Figures	vi
Preface	vii
Acknowledgments	ix
Foreword	xi

CHAPTER 1   ENVIRONMENTAL IMPACT ANALYSIS LAW — 1

Perspectives on Trends of NEPA	2
People-Resources Relationships/Resource-Management Legislation/Evaluation of Water Resource Projects/An Intragovernmental Struggle/Environmental Crises	
The Nature and Content of NEPA	10
The Implementation of NEPA	15
The Interpretation of NEPA	17
The Environmental Impact Statement	21
State and Local Laws Similar to NEPA	23
Planning and NEPA	25
Summary	28

CHAPTER 2   ENVIRONMENTAL IMPACT ANALYSIS — 31

The Interdisciplinary Team	32
Organization for the Analysis	33
Environmental Assessment	35
Preliminary Evaluation/The Environment as a System/Disciplinary Evaluation/An Example/Summary	
Preparation of the Environmental Impact Statement	55
Form/The Question of Size/Substance/Summary	

CHAPTER 3   PUBLIC-PROFESSIONAL RELATIONS — 69

Specialists	69
Officials	76
Citizens	81
Hearings	82

Conclusion	**84**
Bibliography	**85**
Appendixes	**91**
Index	**143**
About the Author	**146**

LIST OF TABLES AND FIGURES

table		page
1	Land Use Spectrum	42
2	Primary Functions of Selected Agencies	78

figure		
1	Environmental Impact Analysis Flow Chart	14
2	Sample Job Sheet Plan for a Private Firm	34
3	Environmental Impact Analysis: Flow Chart and Check List	36
4	Oswego, New York and Vicinity	54
5	Environmental Impact Assessment: A Diagrammatic Representation	55
6	Square of Opposition in Formal Logic	61
7	Extremes of Organizational Location for the Office of Environmental Analysis	72
8	Typical Flow of Activities in the Preparation of an Environmental Impact Statement	77

Preface

Successful environmental impact analysis is an art. Compliance with both the spirit and the detail of environmental law must be blended with an awareness of the extent, theory, and application of scientific concepts of environmental assessment and with effective reporting of that assessment to the public and decision makers. Thus the organization of this book.

Chapter 1 treats the formal, legal basis of environmental impact analysis, the National Environmental Policy Act (NEPA). The section includes consideration of NEPA's origins, its implementation and interpretation, how its utility is focused in the environmental impact statement, and how NEPA has served as a model for state and local governments.

Chapter 2 examines some broad aspects of environmental impact analysis. It is not concerned with the basic principles, biases, and theories of the specialists on the interdisciplinary team that conducts the environmental impact analysis. These individual disciplines are well presented in a number of disciplinary texts, handbooks, and reference manuals, which need not be summarized here. The focus is rather on the processes through which the disciplines can best interact and organize the impact analysis; the environmental assessment that constitutes the bulk of the analysis procedure; and the means of effectively preparing, compiling, and writing the environmental impact statement that reports the analysis.

Chapter 3 deals with people. Environmental impact analysis poses a delicate situation: since its use is predicated upon the need for a reasonable restraint on development, environmental impact analysis represents a means of simultaneously evaluating and controlling the quality of the human environment. The job is done for people and, of necessity, by people. It is therefore important to dwell on such topics as the make-up of the team that conducts the analysis and the relations of the team either with superior units of the organization in which it is located or between the sponsor of a proposed action and consultants. The roles of public officials and citizens at large are considered in light of the particular problems, opportunities, and challenges posed by environmental impact analysis. The purpose of public hearings and the roles of specialists, officials, and citizens in the hearing process is an important and concluding concern.

The basic philosophy behind this book is that the environmental impact statement process is a beneficial one; it is cost effective and improves the quality of the human environment, both spiritually and materially. It is not a device to preclude development, but rather a tool that should be adopted at all levels of human endeavor wherever people and their environments interact. As a basic planning tool, whether or not it is required by federal, state, or local laws, the environmental impact statement process is a sound and responsible undertaking.

Overall, the book is designed to provide the reader with the wherewithal to participate in environmental impact analysis, whether as a member of a team conducting the analysis, as a public official complying with the legal requirements, or as an interested citizen. By itself, it is not the primary source of scientific material for the environmental specialist, nor is it a manual of procedures for the government official or a standard of living guide for the champion of public rights. Environmental impact analysis is basically an exercise in communication, and this book intends to start that process. If it improves that communication, it will have achieved its purpose.

Acknowledgments

I gratefully acknowledge the help of a large number of people who have contributed in a variety of ways to this volume.

I am particularly grateful to two of my colleagues and friends. Lee P. Herrington of the State University of New York College of Environmental Science and Forestry (CESF) offered continued encouragement and ideas, co-authored an early version of this book, and contributed to the college's course in environmental impact analysis on which we collaborated; Sidney L. Manes, Esq., reviewed portions of the manuscript and has always been ready to answer legal questions and to provide encouragement and assistance.

In an effort to up-date and clarify information in the book, I had the good fortune to be able to meet with several government officials who took the time from busy schedules to answer questions and to provide information and insights. This was particularly valuable at a time when the budget cuts of the Reagan administration threaten to undermine or eliminate the regulations and administering agencies for programs to which this book is addressed. The survival of the EIS process in spite of short-sighted budget cutting will be a testimonial to these and other individuals who devoted time and energy to creating and developing it. To Gerald D. Seinwill, Acting Director of the Water Resources Council, William D. Dickerson, Director of the Resource Development Division in EPA's Office of Federal Activities, Peter F. Smith, Assistant Director for Air, Water, and Solid Waste in USDA's Office of Environmental Quality, and Robert B. Smythe, Senior Staff Member of the Council on Environmental Quality, I am deeply indebted.

Peter F. Smith of EPA, George Stauffer of CESF, and former Commissioner of New York State's Department of Environmental Conservation Peter A. A. Berle reviewed earlier manuscript drafts and gave me much valuable criticism and many suggestions. Jeri Lynn Smith of CESF provided editorial assistance on material that was incorporated into the final draft.

Much of what I know of environmental impact analysis has come from the review of others' work and from individuals who graciously gave their time and effort to help make our course and my professional activity in environmental impact analysis a success. The list is lengthy: Hugh O. Canham, Richard I. Goldsmith of Syracuse University, Richard W. Lalor, Phillip J. Craul, David B. Harper, C. W. A. Macey, Arthur R. Eschner, Robert D. Hennigan, Kevin T. McLaughlin of the Power Authority of the State of New York, Cordon C. DeAngelo of the New York State Department of Transportation, Allen F. Horn, Jane Magee of the New York State Department of Environmental Conservation, Robert C. Morris of the Central New York Regional Planning and Development Board, Robert DeSyn of the Rochester Electric and Gas Company, James P. Karp of Syracuse University, Norman A. Richards, John P. Felleman, Richard V. Lea, David T. Shub of the New York State Department of Transportation, Neil I. Gingold of the New York State Department of Environmental Conservation, Allen R. Lewis, and Lawrence W. VanDruff. (Unidentified affiliations are CESF.) Without the assistance of these fine professionals, this book would not have been possible.

My serverest critics have been the many students who have enrolled in our environmental impact course; they continually raised questions about the EIS process itself and my interpretation of it, thus challenging me to present the information in a better and more comprehensible manner. Their contribution to this volume is incalculable, and I am deeply grateful to them.

Nevertheless, as is the case for the leader of a team that prepares an environmental impact statement, responsibility for errors of commission and omission is mine. And as is also the case for the EIS project manager or team leader, I welcome having any such errors called to my attention.

Foreword

It has been more than ten years since passage of the National Environmental Policy Act with its requirements for preparation of environmental impact statements for major federal actions. During this period, significant progress has been made in the preparation of these statements and assessments. The initial inadequate two or three page environmental impact statements justifying proposed federal action and the later telephone directory sized statements detailing thousands of individual plant and animal species have generally been replaced with much more useful documents.

Today the NEPA process has become an integral part of the planning process in many federal agencies. For example, the NEPA process is thoroughly integrated into the Principles and Standards for Planning Water and Related Land Resources as promulgated by the United States Water Resources Council. The revised *Principles and Standards* are a set of procedures used by major federal agencies (the U.S. Army Corps of Engineers, the Soil Conservation Service, and the Bureau of Reclamation for evaluating and planning major federal water projects and programs. The incorporation of the NEPA process into the *Principles and Standards* has reduced duplication while insuring that the evaluation of environmental quality effects will proceed hand in hand with evaluation of economic benefits and costs.

Perhaps the most worthwhile aspect of the NEPA process is that it facilitates communication between individuals and organizations concerned about a proposed project or policy. By providing a common data base and range of alternatives, the NEPA process facilitates rational discussion and analysis between project opponents and supporters. Moreover, the data base and analyses contained in environmental impact statements or environmental assessments often find use beyond the original project or policy and may be of significant use for assessing the environmental effects of similar projects or policies in areas that do not themselves require the NEPA process. The NEPA process is a procedural one; the analysis, conclusions, and recommendations are not binding on the federal agency. Despite this limitation, however, the NEPA process has proved its worth by facilitating informed communication between those agencies, private entities, and individuals concerned about or involved in a proposed federal action.

Preparing an environmental impact statement or an environmental assessment that will be a useful part of the planning and decision process is a difficult task. Many problems and obstacles confront the preparation of a useful environmental impact statement. Imperfectly understood natural processes and systems, lack of adequate data, stochastic processes, and difficulties and uncertainties in forecasting future demand are only a few of such problems and obstacles. Perhaps the most difficult aspect of preparing an adequate environmental impact statement is the bringing together of the many disparate disciplines involved.

In writing this book, Professor Black has made an important contribution to insuring the adequacy of the future NEPA process. He concentrates on many of the difficult aspects of preparing a NEPA statement: bringing together the many necessary disciplines, writing an organized environmental assessment or environmental impact statement so that it is a useful document, and organizing the team that will prepare this document. He devotes extensive analysis to the importance of the public hearing and the public information process. He is to be commended in preparing a book that attempts to bridge the gap between the many disciplines involved in preparing an environmental impact statement in order to insure a useful product.

Professor Black makes an important contribution to the education of a new generation of engineers, scientists, political specialists, and technicians who will carry out the NEPA process in the future. I hope that there will be a NEPA process for them to carry out. The National Environmental Policy Act and its attendant requirements for environmental impact statements have long been the targets of criticism and opposition from many quarters. Some of this criticism is deserved and revisions have been made. With the change of administrations in Washington, however, there exists justifiable concern over the continuation of a rational and efficient NEPA process. I hope that these fears will not materialize and that Professor Black's book will continue to have an audience for years to come.

Leo M. Eisel, Ph.D.
Former Director,
U.S. Water Resources Council

Environmental Impact Analysis

CHAPTER ONE

Environmental Impact Analysis Law

The National Environmental Policy Act (NEPA) (42 USC 4321) was passed by Congress in 1969 and signed by President Nixon on January 1, 1970 as the first federal act of the "environmental decade." NEPA is a recent step in a long series of developmental, cultural, legal, economic, governmental, and often unplanned (or crisis-reaction) efforts by the American people to manage effectively their natural resources and maintain the quality of the human environment.

In a historical perspective, NEPA is a logical step in the long struggle between the champions of both the physical environment (air, land, and water resources) and the intricately linked life forms that constitute the biological environment on the one hand and the leaders of economic development on the other.

The coincidence of America's expansion with the Industrial Revolution spawned widespread resource exploitation, much of which was abusive. As affluence developed from industrialization, the opportunity arose for the introduction of the concept of conservation at the turn of the century. Initially divided into two camps—exploitative and preservationist—the conservation movement now finds a productive focus in NEPA. Differences of opinion concerning the optimum rate of development and the desired extent of preservation persist and reflect the necessity of including considerations of the cultural environment in environmental impact analysis as well as the need for health, aesthetic, and spiritual considerations, especially with regard to pressing urban problems (Smith 1974). "Improved environmental quality and an expanding economy are not incompatible," maintains the Urban Land Institute (1971). "The developer has significant impact on the face of our land and often carries the awesome responsibility of structuring the habitat of people over a long period of time with great investments of capital. He must be drawn into an enlightened program of planning and regulation of our nation's land use." NEPA provides a means to achieve those objectives.

PERSPECTIVES ON TRENDS OF NEPA

People-Resources Relationships

Environmental concerns and resource development philosophies have been expressed by many individuals, several of whom were literary giants as well as observers and interpreters of their environments: George P. Marsh, John W. Powell, John Muir, Gifford Pinchot, Robert Marshall, William Vogt, Fairfield Osborn, Harrison Brown, Aldo Leopold, Barrow Lyons, David C. Coyle, Raymond F. Dasmann, Stewart L. Udall, and Rachel Carson are authors whose words provide a historical perspective on these people-resource relationships.

The highly controversial book by Rachel Carson, *Silent Spring* (1962) "has attained the status of a classic.... If America ever chooses to adopt a sane, coordinated conservation policy—an *environmental* policy—a great deal of the credit must go to Rachel Carson" (Graham 1970). Carson summarizes her message best in the opening and closing paragraphs of the book:

> The history of life on earth has been a history of interaction between living things and their surroundings. To a large extent, the physical form and the habits of the earth's vegetation and its animal life have been molded by the environment. Considering the whole span of earthly time, the opposite effect, in which life actually modifies its surroundings, has been relatively slight. Only within the moment of time represented by the present century has one species—man—acquired significant power to alter the nature of this world.
>
> During the past quarter century this power has not only increased to one of disturbing magnitude but it has changed character....
>
> We stand now where two roads diverge. But unlike the roads in Robert Frost's familiar poem, they are not equally fair. The road we

have long been travelling is deceptively easy, a smooth superhighway on which we progress with great speed, but at its end lies disaster. The other fork of the road—the one "less travelled by"—offers our last, our only chance to reach a destination that assures preservation of our earth.

The choice, after all, is ours to make.

Stewart L. Udall translated and implemented many of the preceeding ideas into governmental action. He invoked his great predecessors in putting a new perspective on the dilemma of exploitation versus preservation. That perspective is the heart and soul of NEPA:

> As George Perkins Marsh pointed out a century ago, greed and short-sightedness are conservation's mortal enemies ... but most of our major problems will not be resolved unless the resource interrelationships are evaluated with an eye on long-term gains and long-term values.... If we are to preserve both the beauty and the bounty of the American earth, it will take thoughtful planning and day-in and day-out effort by business, by government, and by the voluntary organzations.... The conservation movement can become a sustained, systematic effort both to produce and to preserve.... Once we decide that our surroundings need not always be subordinated to payrolls and profits based on short-term considerations, there is hope that we can both reap the bounty of the land and preserve an inspiring environment.... Those who decide must consider immediate needs, compute the values of competing proposals, and keep distance in their eyes as well (Udall 1963).

The last sentence states concisely the dilemma that is to be addressed specifically in each and every environmental impact statement (EIS): the trade-offs between short-term uses and long-term productivity.

Spurred on by these literary and activist leaders, by a host of other dedicated individuals, and by growing public concern, students in colleges and universities provided the prime moving force behind the staging of the first Earth Day on April 22, 1970. That event, in turn, rode the crest of the wave of environmental concern into the environmental decade, "initiated" by the President's signing of NEPA, "the major law to flow from the new 'environmental ethic'" (Hudson 1974), on January 1, 1970.

NEPA brings resource-management decisions and, indeed, any decision that affects resources out into the open, causing the sponsor to disclose the decision-making process, to anticipate consequences, to modify plans so as to minimize environmental damage and, most importantly, to get people to think about the proposed action and to be involved in making decisions. NEPA remains as the only federal law that is comprehensive in its treatment of the environment, in contrast to many piecemeal, often crisis-oriented laws aimed at some limited, although nonetheless important, problem (Costle 1980). NEPA's success may be attributed to the fact that it involves the public; in fact, it depends upon the public for its successful implementation.

Resource-Management Legislation

The philosophies of conservation leaders influenced the development of a body of legislation reflecting the gradual shift from uncontrolled exploitation to responsible management of natural resources (U.S. Forest Service 1952). This continuing shift describes the general history of the principal land-managing agencies (Forest Service, National Park Service, and Bureau of Land Management), the resource-management service agencies (Soil Conservation Service, National Weather Service, and Geological Survey), the two major operational (water-management) agencies (Army Corps of Engineers and the Bureau of Reclamation), and the regulatory agencies (Environmental Protection Agency and the Food and Drug Administration). A great deal of overlap, potential conflict, and no small amount of duplication is inherent in the major missions of these major agencies and many smaller ones. Thus, different agencies share responsibilities for the same resources and activities, such as water-conservation project construction; or they manage different, conflicting portions of the same resource; or they have different functions such as enforcement and regulation, or development and preservation. Such goals may derive from an agency's charter legislation and guarantee continued conflict and misunderstanding within as well as outside the agency.

Generally, the trend over the last 100 years has been to move from single-purpose management to integrated or multiple-use management and, in the absence of successful efforts to house all the resource-managing agencies under one roof, to legislate coordination. The setting of standards is now a more acceptable means for legislation to control environmental quality in areas of specific environmental concern, especially where human health is at peril. What is interesting is that NEPA goes beyond bringing comprehensiveness to our natural resource management, for it includes in our general concern for environmental quality *all* activities of *all* agencies, not just those responsible for resource management activities.

Evaluation of Water Resource Projects

Attempts to evaluate environmental factors draw on the several decades of experience in the area of river basin programs and water resource projects, as well as on the general resources-management legislation. White (1972b) points out that "it should be recalled that the Section 102 (NEPA) type of statement builds upon the experience of 35 years with Federal review of water management projects." The milestones in this history include the 1936 Omnibus Flood Control Act (33 USC 701); the "Proposed Practices for Economic Analysis of River Basin Projects"—which became known as the "Green Book"—and its Bureau of the Budget enforcer, Circular A-47; Senate Document No. 97 of 1962, prepared by the President's Water Resources Council (not to be confused with the Council created by the 1965 Water Resources Planning Act, 70 Stat 244); and the "Principles and Standards for Planning Water and Related Land Resources," adopted in 1973. The progression of planning objectives for water and related land resources projects, plans, and programs through these documents evolved from narrow, economic concerns that exclude aesthetic benefits because they are "intangible" (a self-defeating term) to equalization of environmental and economic values in meeting the objectives of the Water Resources Council's Principles and Standards (1973), which was itself accompanied by an EIS (38 Fed Reg 24779). For a complete analysis of this evolution, see Major (1977). The principles and standards and the accompanying procedures manuals are binding on all agencies. They represent a philosophy of approach to water and related land resources that will continue to evolve in the years ahead, even as budget cutting threatens the WRC.

Black (1975) asserts "that the history of development of [water management] planning objectives leads to the conclusion that economic and environmental considerations should be kept separate":

> The respective mechanisms or tools for critical analysis of each of the objectives already exist: first, the National Economic Development objective is one which may be analyzed, albeit not without difficulty, by the accepted procedures for benefit-cost analysis originally in the Green Book and refined in the later documents which addressed themselves primarily to the wording of objectives, discount rate, and other residual, troublesome but relatively minor problems. And second, the Enhancement of the Quality of the Environment objective should be analyzed by means of the National Environmental Policy Act.... The separation is implied in 102(2)B in that "all agencies of Federal Government shall identify and develop methods and procedures ... which will insure that presently *unquantified* environmental amenities and values may be given appropriate consideration *along with* eco-

nomic and technical consideration." In spite of the foregoing, each analysis should include reference to or a brief, non-technical summary of the other, for purposes of enlightening the reader and maintaining proper perspective [emphasis added].

As one would expect, the evolution of the planning objectives is directly associated with that of resource management legislation, and it parallels the power struggle with the federal government.

An Intragovernmental Struggle

The enactment of NEPA may be thought of, too, as the most recent in a long series of steps in a continuing power struggle between the executive and legislative branches of the federal government. This struggle surfaced as a result of overlapping responsibilities in water resource development by two agencies. Under provisions of the 1936 Omnibus Flood Control Act, the activity of the Corps of Engineers extended upstream, eventually into conflict with Bureau of Reclamation on the Missouri River. The Bureau had been extending its influence downstream under its charter, the 1902 Reclamation Act (33 Stat 388, 43 USC 371), to provide irrigation water to arid lands and to include hydropower production and, of necessity, flood control whenever building dams. Under pressure from Congress to resolve agency differences about projects on the Missouri River, the Departments of Agriculture, Army, Commerce, and Interior, and the Federal Power Commission established the informal, coordinating Federal Inter-Agency River Basin Committee in 1943. Acting either too weakly or too late (and perhaps both) to satisfy legislators and their need, in turn, to satisfy their constituents, Congress officiated at the "shotgun marriage" of the Corps and the bureau on the Missouri by dividing the pork-barrel pie for them: the action is known as the "Pick-Sloan Plan," and was actually initiated when a war-strained President Roosevelt ordered the Corps' Colonel Lewis A. Pick and the Bureau's W. G. Sloan to arrive at a compromise on their overlapping plans for dams and other works on the river (Coyle 1957). Since the agencies depended upon Congress for funds, they could only comply. The "plan" was formalized by Congress in the Flood Control Act of 1944 (58 Stat 887). As is still the case, the Congress preferred not to exert any more pressure than is necessary to get its goals (or those of its constituents) achieved; thus, the Pick-Sloan Plan is the first of many "nudges"—one part of a series of actions and counteractions—that mark this continuing battle between the executive and the legislative branches.

The power of the executive to manage natural resources effectively was weakened further in 1943 by President Roosevelt's loss of his "long struggle to obtain permanent status for a national resources planning organization . . . [which] was accomplished, in part, by the efforts of the bi-partisan, congressional rivers and harbors bloc, who wished to see the Corps put in charge of preparing advance plans for post war projects" (Holmes 1972). In 1946, Congress attempted to control the executive's activity in resources management further by enacting the Fish and Wildlife Coordination Act (60 Stat 855) and the Administrative Procedures Act (60 Stat 237). The former required dam-sponsoring agencies particularly to notify the Fish and Wildlife Service and later, when the act was amended in 1958 (72 Stat 563), to wait for comments on the proposal. The latter standardized procedures for rule making and adjudication and became a means of securing NEPA's effectiveness (see pp. 17-21).

The next step in the see-saw battle between the Branches was the adoption of the Green Book (Subcommittee on Benefits and Costs 1950), which detailed the procedures that the agencies were to follow in the evaluation of river basin projects. The document provided the details necessary to implement the benefit-cost analysis pioneered in the 1936 Omnibus Flood Control Act's general language. More or less simultaneously, through the first Hoover Commission, the executive sought to arrive at some suggestions for self-control in 1949. One of the commission's recommendations was to place all natural resource agencies under one roof, an almost perennial request by some group or other on the Washington scene ever since. But, the cumbersome congressional committee system (Schad 1968) and the seniority system, which perpetuate old ways and power (Green et al., 1972), have been stout obstacles to this generally desirable modernization, although some of the old guard power has been eroded since 1968 by internal congressional reform and by the executive's taking some of the wind out of its sails by Reorganization Plans 3 and 4 of 1970. In these, President Nixon combined certain agencies into the Environmental Protection Agency and the National Oceanic and Atmospheric Administration.

The controversy of the early 1950s over building Echo Park Dam on the Green River in Utah, along with a gradually awakening public that had more and more opportunity to visit and be concerned about national parks and forests, pressured Congress to amend and strengthen the Fish and Wildlife Coordination Act (see above). The Echo Park controversy produced a victory for the preservationists, made organizations like the Sierra Club national in scope, and shifted the location for a major regulating dam to Glen Canyon on the Colorado River. It also resulted in the creation of the President's Water Resources Council, a temporary group that was an updated version of the Federal Inter-Agency River Basin Committee. It is interesting to note that Congress upstaged the executive by publishing the President's Water Resources Council's "Policies, Standards, and Procedures . . ." as a Senate document (Water Resources Council 1962).

The 1965 Water Resources Planning Act (79 Stat 244), the 1967 Freedom of Information Act (8 Stat 54), and the 1968 Intergovernmental Coordination Act (82 Stat 1098) were increasingly forceful restrictions on the executive branch. As a direct consequence of the Planning Act, the permanent Water Resources Council (which it created) proposed new principles and standards in 1970, and the president did, by the reorganizations plans mentioned above, commence a more sensible structure of the executive branch with regard to environmental quality and pollution-control program agencies. But public pressure and the intermittent, short steps by the executive branch were insufficient, and Congress loaded the legislative cannon with NEPA. The act "has led to significant procedural reforms" (Carter 1976), and began a revolution within the entire federal government, for the courts were compelled to enter the continuing struggle for power by arbitrating between the other two branches.

The specific history of NEPA in the waning days of the 1969 session of the ninety-first Congress is traced in detail by Anderson (1973). Historians suggest that NEPA started with two 1968 reports, one by the Subcommittee on Science, Research, and Development, and the other by the Senate Committee on Interior and Insular Affairs. But it is safe to trace NEPA all the way back to the original Fish and Wildlife Coordination Act, for it contained a last-minute amendment by Congressman Aspinall to extend the EIS process to all impacts, not just fish and wildlife. Politically, the act became a focus of a jurisdictional struggle between two presidential contenders, Senators Jackson and Muskie, whose respective Senate committees (Interior and Insular Affairs, and Air and Water Pollution Control, a subcommittee of the Committee on Public Works) each had legitimate interest in the bill.

Environmental Crises

Increasingly serious air pollution events during the last 100 years in the United States and England have given strong impetus to the need for effective programs of environmental quality control. It has become imperative to see the link between air, soil, and water pollution as well as problems of resource management: to be comprehensive in scope. Although air pollution problems have a history documented back to the eighteenth century, serious pollution episodes and situations have been increasing both in frequency and in magnitude, most significantly during the growth period of the industrial world. Pollution problems differ as to their noticeability and permanence, depending on their relation to affecting factors in both time and space. The increasingly unhealthy state of affairs resulting from such pollution has resulted in strong public support for effective programs for environmental management. Air pollution episodes, as they are called, are transitory, that is, they come and go with the weather. But they cannot be ignored when they strike. They can be deadly. The famous London black fog hung over that city in 1952 for four days and killed 4000 people. That particular episode triggered public action and legislation that resulted in the long-term cleansing of the London atmosphere.

Although the United States has had its share of similar episodes, such as those at Donora, Pennsylvania in 1948 and New York City in 1966, none have been as severe as the London fog. Of more concern has been the gradual degradation of the atmosphere of the entire nation by the automobile (Esposito 1970). The Environmental Protection Agency estimates that "air pollution costs the people of the United States an estimated $20 billion every year" (Environmental Protection Agency 1976).

The Dust Bowl of the 1930s led directly to the formation of the Soil Conservation (originally Erosion) Service and other agencies and to the adoption of widespread soil conservation practices. Thomas Jefferson valued the link between soil and vegetation conservation in his concept of a small-holding, agrarian democracy (Moncrief 1972). Implementation of the concept by the federal agencies acquired legal acceptability via the commerce clause of the Constitution through control over navigable waters, as is apparent in the 1899 Refuse Act (30 Stat 1152), actually a rider to that year's Rivers and Harbors Act; it assigned to the Corps of Engineers, in recognition of its authority to control and develop navigation, the responsibility of issuing permits for deposition of any substance in navigable waters. No machinery for administering the permits was established (Holmes 1979).

The war against pollution of waterways and bodies of water began in earnest in 1948 and has been different from that against air pollution; the ecological alteration (and, more rarely, the destruction) of the aquatic ecosystem may be slow and, once wrought, is often even slower to repair. Creation of an effective set of institutions to combat pollution is also difficult and slow.

The scattered documentation available at the time on the importance of water pollution and food chains was drawn together in *Silent Spring* by Rachel Carson in 1962. Gradual deterioration of water was observed in estuarine ecosystems, with the loss of commercial shell fisheries, fish kills along the Mississippi River, the gross pollution of Lake Erie and other lakes, and the simultaneous recognition of the impact of concentration of DDT derivatives through aquatic food chains (Wurster and Wingate 1968). The environmental destruction that accompanied the Vietnam War also came to the attention of the public and played a role in the attitudes of the times.

All of these environmental crises contributed to the public awareness and led, in part, to NEPA. Perhaps the final inputs were the view from space by the early astronauts, who saw forest fires from their spacecraft, and the high-quality photography that brought the environmental message even more directly to the population via color television (Barbaro and Cross 1973). Concurrently with these crises, technological advances prompted improvements in instrumentation to monitor environmental conditions and to evaluate civilization's overt and inadvertant impacts (Changnon 1976). The development of reliable field and laboratory equipment, computer analysis of intricate environmental factors, and large masses of data, along with greater understanding of complex environmental processes, permit prediction of impact as well.

THE NATURE AND CONTENT OF NEPA

NEPA consists of both a statement of national environmental policy and an effecting mechanism. Its brief text contains three general parts: a declaration of policy and an action-forcing section, both in Title I, and the creation of the Council on Environmental Quality (CEQ) in Title II.

NEPA's purpose is best expressed in the language of Section 2 where it states that "the purposes of this Act are: to declare a national policy which will encourage productive and enjoyable harmony between man and his environment; to promote efforts which will prevent or eliminate damage ... ; to enrich the understanding of the ecological systems and natural resources ... ; and to establish a Council on Environmental Quality" (see Appendix A for full text). Title I of the Act declares the policy and spirit of the law:

> The Congress, recognizing the profound impact of man's activity on the interrelations of all components of the natural environment ... declares that it is the continuing policy of the Federal

Government ... to create and maintain conditions under which man and nature can exist in productive harmony.

This does not create for the citizens of the United States a substantive right to some specified condition of the environment, for the act utilizes the word *should* rather than "shall" with respect to individual rights to a healthful environment and attendant responsibilities.

The action-forcing section demands that any action undertaken by a federal agency be preceded by 1) close examination of the environmental consequences of the action (the assessment), 2) full and public disclosure of that examination in the draft environmental impact statement (DEIS), and 3) review by any and all interested parties, which must be summarized, compiled, published, and responded to by the agency in the final environmental impact statement (FEIS). Specifically, Section 102 declares that all units of the federal government shall interpret and administer their activities in light of Section 101 and, in the best-known part of the act, states:

(2) all agencies of the Federal Government shall ...
   (C) include in every recommendation or report on proposals for legislation and other major Federal actions significantly affecting the quality of the human environment, a detailed statement by the responsible official on
   (i)   the environmental impact of the proposed action,
   (ii) any adverse environmental effects which cannot be avoided ...
   (iii) alternatives to the proposed action,
   (iv) the relationship between local short-term uses and long-term productivity, and
   (v)   any irreversible and irretrievable commitments of resources.

Conspicuous by its absence from Section 102(2)C is any mention of who enforces it. Enforcement is actually left up to the public, from which an important principle about NEPA may be inferred: if the public doesn't care, no one else will. The role of the Environmental Protection Agency (EPA) in enforcement of related environmental legislation is discussed in the section on Implementation of NEPA (pages 15-17). EPA's authority for this activity is established by Section 309 of the Clean Air Act (42 USC 7609) of 1970 (see Appendix D).

Further sections of Title I cover broad details of responsibility and implementation, as well as some specific goals relating to environmental quality, resources management, and human involvement with them.

NEPA formalizes environmental impact analysis, the basis for evaluation of civilization's effects on environmental quality. Environmental impact analysis consists of (1) determination of the components and interdependencies of the environment potentially affected by a proposed action, (2) assessment of environmental quality with and without the proposed action and its alternatives, (3) preparation of an environmental impact statement that reports the findings of the determination and the assessment, and (4) review of the DEIS by all legitimately interested publics and governmental units and compilation into an FEIS. The EIS is, therefore, at once the evidence of compliance with both the procedural content of the Section 102 requirements and the substantive content of the Section 101 goals.

Title II of the act provides for establishment of a Council on Environmental Quality to advise the president on matters pertaining to the environment, to recommend legislation, and to oversee generally the workings of all that NEPA requires. The council consists of three members, appointed by the president and with the advice and consent of the Senate. The names of the members and their staff, as provided for in NEPA, and the activities of the CEQ may be found in CEQ's annual reports. Some of the studies that the CEQ has undertaken are published separately by the Superintendent of Documents, as are the annual reports. Establishment of the Office of Environment Quality (OEQ), directed by the CEQ and in the Executive Office of the President, is provided for by the Environmental Quality Improvement Act of 1970 (42 USC 4371) (see Appendix B). The Office of Environmental Quality provides the professional and administrative staff for the CEQ. The OEQ and the CEQ have been managed as a separate unit; the CEQ has served as an impartial review mechanism for many federal departmental activities (Baldwin 1981).

NEPA is one of the most fascinating, far-reaching, and controversial laws passed by Congress. It is fascinating because, like the U.S. Constitution, it is short—a little over four pages—yet it has had repercussions throughout the nation's governmental structure. These four short pages have literally caused a revolution in government, creating the CEQ and causing it to create guidelines and the agencies to comply and prepare more than 11,000 EISs in the first decade. NEPA allocates no funds to the preparation of EISs, only for the administration of the CEQ. EISs are to be prepared out of existing budgets: the monies the EIS saves in the long run by an analysis in the early planning stages of an action are expected to outweigh the costs of preparation.

The act is far-reaching because it has been applied to virtually all actions of the federal government, from major construction to the granting of funds or permits, to setting of standards, to planning, and to broad programs such as wilderness preservation and pesticide usage. In addition, NEPA has served as a model for the 25 states, Puerto Rico, and the many local governments that have enacted similar laws to date.

NEPA is controversial for a number of reasons. First, it superimposed a new set of rules on top of existing procedures, a usually unpalatable morsel for any governmental unit. Second, it demands that experts, who have long assumed that all their actions were taken in the public interest and that they—as experts—know best, be considerably more accountable and, in the process, more communicative. Finally, it is controversial because it materialized at a time when inflation and the onset of an energy crisis put uncommon pressures on everyone, making NEPA susceptible to criticism that threatened to erode its effectiveness (Citizens' Advisory Committee on Environmental Quality (1974), not to mention its very existence.

That NEPA has withstood many attacks is testimony to the wisdom of its authors in establishing goals that are lofty yet within reach. The result is a framework for a constructive approach toward maintaining and improving environmental quality, with the impact statement firmly embedded in the planning process. Since the EIS usually precludes costly mistakes and mitigates adverse impacts to the environment, it is a device that pays its way many times over.

In essence, then, NEPA created a process that does not automatically preclude any action expected to have a significant adverse impact on the human environment from taking place. The relationship of this process to some other important elements in the decision-making process is shown in Figure 1. A "no" response to the three questions regarding economics, national security, and political or social viability leads either to a new or modified proposal or to scrapping the original proposal. A "yes" response leads directly to a final decision. Adding the question regarding the applicability of NEPA yields a reversal: a "no" response joins the "yes" responses from the other questions and is the same, essentially, as if NEPA had not been invoked. It follows that, without NEPA, it was possible to take an action without consideration of alternatives, a potentially damaging occurrence when environmental values are considered. With a "yes" response to the NEPA question, however, the consideration of the null and other alternatives (and often a hearing) is mandated if the impact of the proposed action is either significant or controversial. The initial environmental impact assessment (EIA) may be sufficient to document the fact that no significant impact is anticipated or that the action is not controversial, in which case a negative declaration (now, officially, a Finding of No Significant

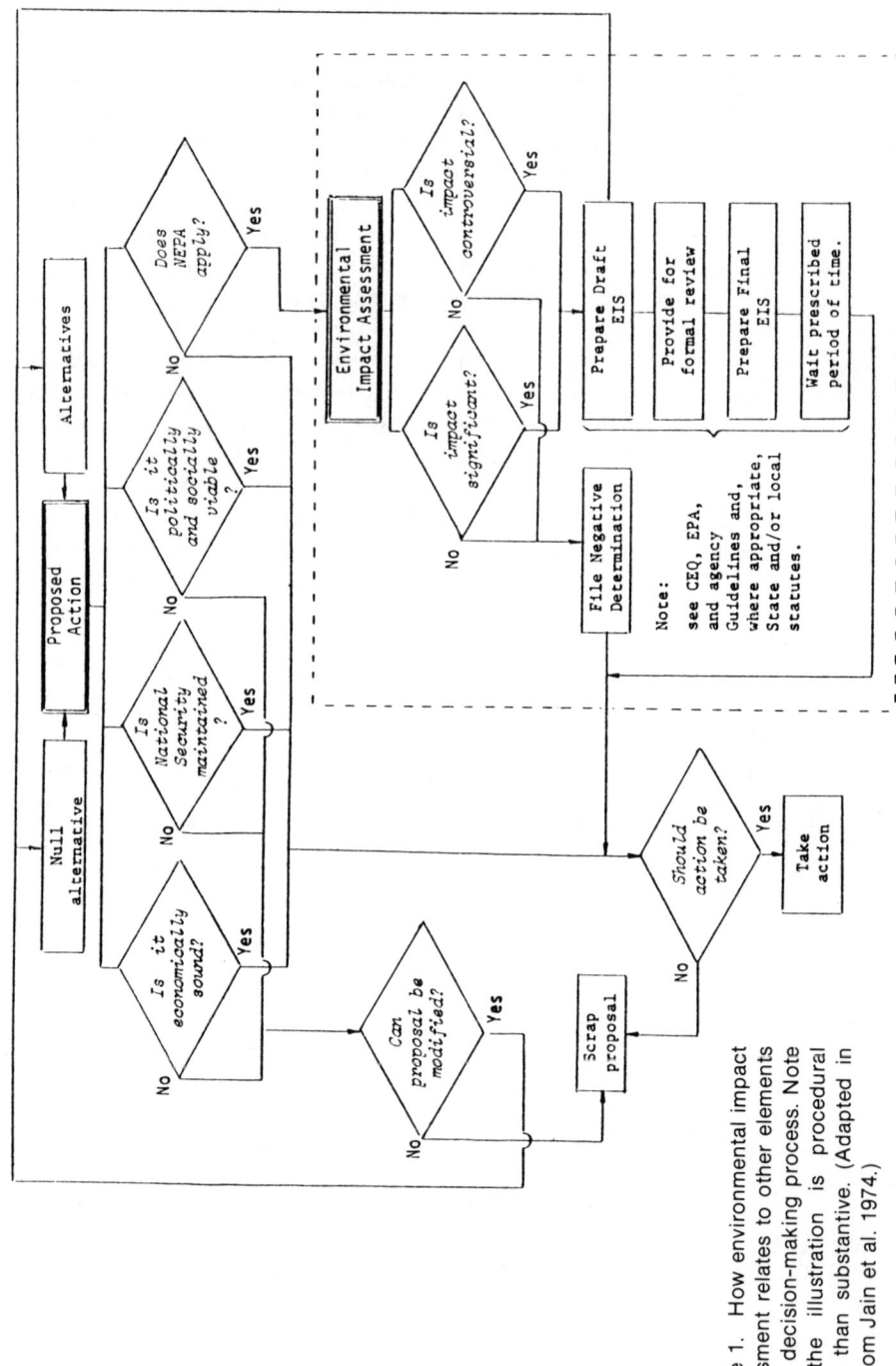

Figure 1. How environmental impact assessment relates to other elements in the decision-making process. Note that the illustration is procedural rather than substantive. (Adapted in part from Jain et al. 1974.)

Impact at the federal level) is filed, and the final decision on the proposal is at hand. The dotted line encloses the entire EIS process, a process that has come to be known as environmental impact analysis: only one point of entry is available, and one of the two points of departure automatically leads back into it. The other exit documents that proper environmental review has taken place and the final decision on the proposal may be made. (Note that simplification of the entire decision-making process in Figure 1 requires omission of similar details of the processes that accompany the other major questions; of course, their omission does not mean they do not exist.)

THE IMPLEMENTATION OF NEPA

To set the provisions of NEPA in motion, and to comply with its own mandate, the CEQ promulgated guidelines in 1970 (35 Fed Reg 7391) that, with the exception of time limits for circulation of the DEIS, left compliance up to the agencies. (As with all other rules proposed by federal agencies, these were subject to review and comment under the Administrative Procedures Act.) With less than a year's experience, the CEQ revised the guidelines and issued new ones on August 23, 1971 (36 Fed Reg 7724). These remained in force until August 1, 1973, when the CEQ issued new guidelines that remained in force until the current regulations (Appendix C) were adopted. By 1976, the agencies had, with the assistance of the CEQ and as required by Section 103 of NEPA, developed their own rules and regulations for compliance with NEPA and the CEQ guidelines (Council on Environmental Quality 1976a).

In his initial environmental message to Congress in 1977, President Carter directed the CEQ "to issue regulations requiring all federal agencies to meet [the] criteria and the provisions of Section 102(2)C" which he accomplished specifically in Executive Order 11514, as amended by Executive Order 11991 on May 24, 1977 (see Appendix E). As a consequence, in June 1978, the CEQ published a completely revised set of proposed regulations, which were finally adopted on November 29, 1978, and became effective on July 30, 1979 (Appendix C). Executive Order 11991 also called for the council to establish procedures that would allow referral of interagency environmental disputes to the CEQ (Baldwin 1981).

In contrast to the earlier advisory guidelines, the new regulations were binding on all the agencies. They provide uniformity between agencies—replacing earlier agency rules (Baldwin 1981)—and introduce some innovations: "Scoping" (Section 1501.7) is introduced at the earliest stages of the EIS process in order to define the issues and procedures specifically needed for the action proposed. The EIS must also contain a "record of decision" (1505.2) that reports on the alternatives considered, summary information on reasons for rejection or modification, and mitigative measures that could be made a part of the proposed action. These measures may become conditions of the approval to commit the action, as is the case in New York State's State Environmental Quality Review Act (or SEQR law), particularly as it applies to the permit process. In fact, both the scoping process and mitigative measures may have derived from the experience in New York, where a "conceptual review" is built into the EIS process and mitigative and ameliorative measures are required inclusions in the EIS. The record of decision also serves as the CEQ's means of enforcing the regulations.

The new regulations also call for a list of preparers (1502.17) and emphasize that the EIS is to be useful and readable by calling for it to be analytic rather than encyclopedic (1502.2), and in plain language (1502.8). A page limitation (1502.7) is set for EISs at 150 (300 pages for proposals of "unusual scope and complexity") and the regulations call for "tiering" (1502.20), that is, preparing EISs for broad programs so that individual project EISs may concentrate on issues germain to that level of decision, as suggested originally by Anderson (1973).

Through the new regulations, NEPA also now calls for better integration with other environmental legislation, including the Fish and Wildlife Coordination Act, the Endangered Species Act (16 USC 1536) of 1976, and the National Historic Preservation Act (16 USC 470) of 1976, and better coordination with the Congress and the states (Council on Environmental Quality 1979). Section 7 of the 1976 Endangered Species Act has "certain procedural resemblances" to NEPA; both habitat and species existence are to be discussed in the EIS (Harrington 1981).

Through Section 309 of the Clean Air Act of 1970, as amended (42 USC 7609; see Appendix D), EPA coordinates EIS review and compliance with environmental standards for which it has responsibility under that legislation, as well as the 1972 Water Pollution Control Amendments (33 USC 1251), the Coastal Zone Management Act of 1972 (16 USC 1451), the Marine Protection, Research and Sanctuaries Act of 1976 (33 USC 1401), the Resource Conservation and Recovery Act of 1976 (42 USC 6901), the Toxic Substances Control Act of 1976 (15 USC 2601), the Safe Drinking Water Act of 1976 (42 USC 300), the Noise Control Act of 1976 (42 USC 4901), and the Federal Insecticide, Fungicide and Rodenticide Act of 1948 as amended in 1978 (92 Stat 827). EPA and the heads of other federal agencies are mandated by Section 309 of the Clean Air Act and by Section 1504 of the new regulations to refer interagency disagreements and environmental problems to the CEQ.

The CEQ saw also fit to focus on alternatives as "the heart of the environmental impact statement" (1502.14), giving this one of the original five inclusions, NEPA, 102(2)C, extra emphasis. The new regulations also call for identification of the "human environment" to be "interpreted comprehensively to include the natural and physical environment and the relationship of people with that environment. . . . [E]conomic or social effects are not intended by themselves to require preparation of an environmental impact statement" (1508.14).

Actions to which NEPA applies are identified in Section 1508.18 in the regulations and are being extended now by Executive Order No. 12114 (Appendix E) to include any federal action that affects the "global commons," namely, the oceans, atmosphere, space, and any foreign country. Definition of the term *action* was one of the more important activities of the courts in their role of clarifying NEPA's languages; along with the many other questions raised about NEPA's applicability, the definition are discussed below.

Finally, the CEQ regulations were cross-referenced to the Water Resources Council's "Environmental Quality Evaluation Procedures" (45 Fed Reg 64446) upon adoption in September of 1980 (Appendix F). At this writing, the procedures are undergoing revision.

INTERPRETATION OF NEPA

While the courts played a role in defining the vague language of NEPA, resulting in part in the current regulations dicussed above, the "enabling" of NEPA really commenced before it was enacted. Two acts and two court cases were particularly instrumental in laying the legal groundwork and the administrative machinery for NEPA and for the interpretation, broad application, and acceptance that was to follow. The first was the Administrative Procedures Act, which established procedures for adjudication and rule making. The act also provided a standard for citizen challenge to those procedures; for example, an agency decision that is made in an "arbitrary and capricious" manner might be contested procedurally. Thus, a plaintiff might successfully contend in court that an administrative decision had been made on grounds other than those that the administrator was mandated to consider. Since NEPA is in large part a procedural act, the "arbitrary and capricious" standard is a viable tool for challenge by the public. NEPA, it should be recalled, did charge "each person with a responsibility to contribute to the preservation and enhancement of the environment" in addition to stating "that each person should enjoy a a healthful environment" (Section 103). Citizens can exercise that responsibility by challenge to agency compliance with the procedures.

Second, the Freedom of Information Act of 1967 started a continuing process of opening up the government (at all levels, for states have enacted similar legislation) to public scrutiny. Decision making, which in many governmental units had withdrawn into the bureaucratic woodwork, was made available to the public for the asking—unless, of course, the information sought was classified. Together, the Administrative Procedures Act, the Freedom of Information Act, and the National Environmental Policy Act provide the means, right, and justification for the public to review and participate in agency decision making. Thus, leaving enforcement of NEPA up to the public was, like NEPA itself, a logical event in a series of improvements in participatory democracy.

In the first of two court cases, *Data Processing Service v. Camp* (397 US 150, 1970), the primary issue was whether damages beyond economic or physical injury might be considered as the basis for standing to sue: the court held that aesthetic, recreational, or conservational injury were grounds in this historic case and, in the second, "Scenic Hudson II" (*Scenic Hudson Preservation Conference v. Federal Power Commission*, 354 F2d 608, 1965, Cert Denied, 407 US 926, 2 ELR 20436, 1972), the court held that the FPC must consider environmental quality over and above its narrow mandate. Both cases are thus responsible for paving the way for major legislation that opened new grounds for citizens to obtain the right to be in court for the purpose of maintaining and enhancing environmental quality.

Thus, in NEPA, the citizen gained standing in procedures, and lack of compliance with those procedures meant that officials might be challenged with arbitrary and capricious decision making. During the early 1970s, the "we-know-what's-best" attitude of many agencies led to their determination to find out how little compliance was needed and what exactly would be required. At the same time, the public was interested in flexing its new found muscle: in testing its ability to slow down—and occasionally to stop—the environmentally heedless agencies. Despite this confrontation, in the first ten years of NEPA's existence, less than 10 percent of the more than 11,000 EISs filed by the federal agencies had been litigated, and injunctions had been issued in only 2 percent of all EISs submitted (Council on Environmental Quality 1979). No doubt a large number of projects have also been scrapped in the early planning stages, owing to agency anticipation of adverse environmental impact or, perhaps more importantly, adverse public reaction.

The landmark cases that dealt with the interpretation of NEPA's vague language and its application are well-documented in the annual reports of the CEQ and by Anderson (1973). Many are known by their popular names which appeared in the media. Although cases are often cited as pertinent to only one issue, they usually dealt with several. Some of the more important cases led to changes in the CEQ guidelines and regulations, as well as in NEPA itself.

Calvert Cliffs (*Calvert Cliffs Coordinating Committee, Inc. v. Atomic Energy Commission*, 404 US 942, 1972) is perhaps the best-known case in that it dealt with a large number of issues, including application of NEPA to ongoing projects, to research and development actions, and to consideration of environmental matters other than the radiological health concerns mandated for the AEC.

Gilham Dam (*Environmental Defense Fund v. Corps of Engineers*, 41 USLW 3637, 1973) dealt with partially completed projects and the requirement for preparation of the EIS by an interdisciplinary team. *Greene County* (*Greene County Planning Board, et al., v. Federal Power Commission*, 455 Fed Rep 412, 1972) treated the meaning of "the responsible federal official." It was found that the commission (which was about to issue a license for a transmission line to which the Greene County Planning Board objected) could not accept the applicant's (Power Authority of the State of New York) environmental package and re-issue it as a valid EIS—that did not constitute preparation of an EIS. "Gulf Oil" (*Natural Resources Defense Council v. Morton*, 337 F Supp 170, 1972) aided in the definition of the word "alternatives," and "Mineral King" (*Sierra Club v. Morton*, 92 S Ct 1361, 1972) treated standing and class actions. *Overton Park* (*Citizens to Preserve Overton Park v. Volpe*, 2 ELR 20501, 1972) was one of the first cases to treat the Department of Transportation's relations with the state departments of transportation insofar as federal funds for state highways were concerned. One of the most recent of these cases is *Conservation Society of Southern Vermont v. Secretary of Transportation* (531 F2d 637, 1976), which clarified the "responsible official" terminology as well as the need for coordination of EISs for federally planned highways and shorter state-constructed segments. This case resulted in the only amendment to NEPA in 1975 (codified as section 102(2)D of NEPA). "Amchitka" (*Committee for Nuclear Responsibility v. Seaborg*, 1 ELR 20469, 1971) resulted in the clear requirement that the final EIS must include not only the comments submitted by interested parties on the DEIS but an analysis of and response to them as well.

From these and other landmark decisions, two facts became abundantly clear. First, once the EIS requirement was met procedurally by an agency, the action could proceed. Thus, if there is interest in stopping a proposed action, a group, individual, or the public at large is more likely to be successful if grounds other than NEPA are invoked. NEPA is not designed to stop development, only to insure that whatever development takes place protects, maintains, or enhances environmental quality.

Second, although many argue that EISs have been rejected on substantive grounds (see, for example, Council on Environmental Quality 1976b, p. 404)—that is, that they are inadequate in their coverage, analysis, or application—legal grounds are more easily secured on procedural bases, again leading to the conclusion that, once the procedures have been satisfactorily complied with, the action may be executed. Clearly, if the EIS lacks substance, one must conclude that it was inadequately prepared. The point may well be moot, since the evidence that the act has not been complied with procedurally is usually substantive; in either case, once the agency complies the action may proceed.

After numerous court cases and administrative decisions, several types of actions have emerged as exemptions to those requiring EISs. Emergency and temporary actions are generally exempted, as are classified actions in which full disclosure would obviously not be in the national interest. Revenue sharing is also specifically exempted (1508.18) as are "categorical exclusions" identified by specific agencies in cooperation with the CEQ. In addition, Congress specifically excluded certain activities of the Environmental Protection Agency from the provisions of NEPA, but the EPA has prepared EISs on several regulatory actions on a voluntary basis and will continue to do so (44 Fed Reg 64174, November 6, 1979). EPA is required to prepare EISs for permits for new point sources of pollution for which performance standards have been set and for grants for construction of new waste treatment facilities.

Two recent decisions provide material that may be incorporated in future revisions of or to the CEQ regulations. In RARE II (Roadless Area Review and Evaluation), a programmatic EIS that identifies about 3000 tracts in 38 states (approximately 62,000 acres) for wilderness designation, the court identified a lack of site-specific data as a deficiency to an EIS:

> Site-specific analysis is essential to meaningful environmental analysis. Broad, generic statements neither inform the public of the environmental consequences of the action, nor require the agency

to take a "hard look" at environmental factors.... The Forset Service's justification for the failure to provide detailed site-specific examination of the areas in question is bottomed on the notion that compliance with its statutory NEPA duties would render the environmental statement too bulky. A statutory duty cannot be excused simply because it is too difficult to perform. (*California v. Bergland* 483 FSupp 465, 1980; quoted in Society of American Foresters 1980).

In a similar vein, but applying to overall alternatives to a proposed action rather than to multiple applications of a broad program, the court ruled in *Grazing Fields Farms v. Neil Goldschmidt* (F2d, 1st Cir. 1980) that the alternatives must be discussed in the EIS, not merely referred to in an "administrative record." Specifically, the court held that pertinent information found in the administrative record but not incorporated into an EIS cannot satisfy the statutory requirement to discuss alternatives. This ruling may affect other uses of the concept of incorporation by reference (Section 1502.21 of the 1979 CEQ regulations) as well. Clearly, the interpretation of NEPA and of the EISs it spawned is a continuing process involving the courts, and the agencies, and the public.

THE ENVIRONMENTAL IMPACT STATEMENT

What *is* an EIS? Actually, the term is not readily defined, although each of the individual words is defined in section 1508.1 of the new CEQ regulations. It is, as stated above, more of a process than anything else. But the focus is on draft or final document that reports existing environmental conditions, the proposed action, and the expected interaction between the two. The EIS is a "full-disclosure" document; that is, it is supposed to present a record of all the decisions regarding the environment-action interaction, as well as the assessment thereof. The difference between the DEIS and FEIS is that the latter reports the sponsoring agency's summary and reaction to the public's and other agencies' comments on the DEIS. Both may include *mitigative* measures, which will reduce the severity of impact or *ameliorative* measures, which will make the impact positive so that the environment would be improved. As a decision-making tool, the EIS takes its place alongside several other aids that should be available to the decision maker, such as the benefit-cost analysis and other analytical reports on a proposed action. Ideally, the decision maker should have all these reports available to make a reasoned, intelligent, acceptable, and often justifiable choice between alternatives. Thus, the EIS preparers must compete with other, sometimes emotional or simplistic tools for the attention of the administrator, politician, or citizen.

The purpose of the EIS is, therefore, to embody the "practicable means" whereby "Federal plans, functions, programs, and resources [are] improved and evaluate[d]" so as "to promote efforts which will prevent or eliminate damage to the environment and the biosphere and stimulate the health and welfare of man" (NEPA, Section 2).

Generally, the objectives of the EIS are to satisfy the law, to evaluate environmental impact, and to improve decision making throughout the federal bureaucracy for virtually all actions. To accomplish the first objective, an agency must comply with its own rules and regulations and with the CEQ regulations, as well as with the letter and spirit of NEPA. Evaluation of impact entails comparing the expected environmental consequences of several alternatives to assist in selection of that alternative which contributes most to society's goals, including consideration of the minimization of environmental damage. Improved decision making means the development and presentation of facts upon which the decision is based. The criteria for the development of facts include relevance, perspective, and basic precepts of the scientific method as applied to environmental impact investigations (not necessarily the same as those of "pure" or basic research efforts, but with the same degree of objectivity). The criteria for the presentation of the facts include deduction, brevity, utility, and documentation. More specifically, the goals of the EIS are to:

1. Identify hazards of action and alternatives
2. Evaluate environmental benefits and costs
3. Suggest mitigative and ameliorative measures
4. Notify legitimately interested publics
5. Inform the public at large
6. Provide useful information to decision makers

The first three goals require that environmental impact analysis be conducted and that the EIS be prepared by a disinterested, professional team. NEPA specifies, and the courts have upheld, that the EIS shall be prepared by use of "a systematic, interdisciplinary approach" (Section 102). Recognizing the need for expertise in order to fully appreciate, observe, and interpret environmental impact on a specific portion of the environment, Luken and Langlois (1973) observe that the team approach is necessary "to correct the biases of technical intelligence." Further, most of our environmental problems have not arisen through single-handed or simple action; nor can their successful resolution be simple minded. Thus, the University of Wisconsin's Leonardo Symposium on Resources and Decisions (Leonardo Scholars 1975) invoked a primary environmental truth in urging the use of interdisciplinary teams of experts: "in diversity lies the basis for survival."

The last three goals require that environmental impact analysis be placed on a par with benefit-cost analysis, which normally deals exclusively with economic impacts. Both should commence in the early planning stages of project development, but both actually are processes that continue throughout the planning, design, and construction phases of a given action. As suggested earlier, the environmental impact analysis, represented by the EIS, is just one of the many documents upon which a decision is based. Thus, the environmental scientist, as a member of a team, is challenged to present the EIS in a succinct and effective manner.

Finally, the role of the public in the EIS process is not quite as explicit as that of the experts and the agency decision makers. In a democratic society, where public opinion both directly and indirectly influences the decisions that are made, it is essential that the public be honestly informed about the activities of government. Furthermore, while there are no sanctions contained within NEPA for failure to comply with its provisions, it is incumbent upon the public to make it work. The law spells out what the public servants will do in order to comply; the law also challenges the public to play a major, constructive role.

STATE AND LOCAL LAWS SIMILAR TO NEPA

The passage of the National Environmental Policy Act has had ramifications in state and local laws, and in citizen participation with regard to protecting and enhancing the quality of the human environment. A large number of the states and innumerable other governments have enacted legislation that parallels NEPA in spirit and often in extent of application, although these laws usually apply only to actions for which a federal EIS is not required. In addition, NEPA has given citizens the means to enter the decision-making process at all levels with much greater ease than was previously the case.

Fifteen states and the Commonwealth of Puerto Rico have adopted formal requirements for environmental impact analysis and EIS preparation similar to NEPA by legislative action. One of these, New Mexico, repealed its "little NEPA." Eleven other states have administrative or limited versions. The fourteen states with "little

NEPAs" still in force are California, Connecticut, Hawaii, Indiana, Maryland, Massachusetts, Minnesota, Montana, New York, North Carolina, South Dakota, Virginia, Washington, and Wisconsin. Arizona, Michigan, Nebraska, New Jersey, Texas, and Utah are the states with administrative NEPAs; Delaware, Georgia, Mississippi, Nevada, and North Dakota have NEPA-like legislation that applies to certain specific types of actions, which varies with the state. Other than the applicability, the EIS requirements for the states generally differ little from the federal model (Trzyna 1974).

There are some differences between states as to the application of requirements to private activities and local government actions, with varying threshold levels. Of course, many state (and some private) actions involve obtaining federal permits, licenses, cooperation, or funds, so NEPA itself may govern the action at hand. For actions that require a federal EIS, the state law specifically eliminates duplication. Thus, it is expected that the state act is most effective (and most objectionable to local taxpayers) when it applies to private and/or local projects.

State laws may have built-in approval points in the EIS process, unlike the federal model. An example in New York State is that the procedural clock of a permit-granting agency, for example, is started (specifying timing of hearings, opinions, and so on) only after acceptance of the DEIS. That acceptance constitutes judgment by an agency that the EIS is complete, a decision point that is not inherent in the NEPA process. Similarly, New York invokes the EIS process whenever two or more permits are required in order to complete an action and combines the hearings; when this occurs the EIS process, which ideally is supposed to be full disclosure in nature and not subject to a "right" or "wrong" type of decision, is intimately linked with the permit process, which inherently has a "yes" or "no" conclusion associated with it: the two are not necessarily compatible (Manes 1980).

Few states that have strong land-use planning laws have EIS requirements, although Hawaii has both. Several other "little NEPAs" require discussion of the relation of expected impact to existing land use plans, as required in the 1979 CEQ regulations (1502.16c) and in earlier guidelines.

Enforcement is not treated in any of the state acts (excepting Minnesota) which, therefore, like NEPA, rely on the full-disclosure concept and citizen participation. Nevertheless, if a permit for private development depends upon submission of an EIS, a regulatory agency obtains enforcement power *a priori*. All of the states have created or designated a specific agency to administer the act;

but without sanctions for noncompliance, the burden of enforcement ultimately falls upon an alert citizenry.

Many lower governmental levels have adopted NEPA concepts, as evidenced by responses to a CEQ questionnaire which revealed "that the benefits of the EIS process to decision-making outweigh the administrative burdens and uncertainties" (Council on Environmental Quality 1979). Clearly, the planning benefits of NEPA and its offspring are attractive in the long run.

PLANNING AND NEPA

A plan is a scheme of action, a series of steps that are to be undertaken to achieve some specified goal dictated, presumably, by some policy. Certainly, it would seem that planning is an essential ingredient in the game of survival. Yet, "in the United States, planning is suspect" (Spilhaus 1972); worse still, the establishment of an overall policy as a planning base is anaethema to the "free enterprise system." However, farmers, contractors, and independent professionals—all people in business—plan. So do the agencies of government, but not always in a concerted manner, nor on the national level. The argument is with comprehensive plans, as opposed to fragmented or eclectic plans: individual, corporate, and even governmental lobbyists bring pressures to bear at concentrations of power and seek either to restrict that power or to bend it to self-serving ends. As a result, attempts at comprehensive planning are often less in the interests of unrestricted freedom of the firm and the individual than is heralded. NEPA, and the EIS process it spawned, flies in the face of those who seek to minimize formal, open, comprehensive planning.

The reason behind this conclusion is simply contained in the statement of purpose of NEPA: that section deals with goals that can be achieved only over a long period of time—with planning. The responsibility is left to both the preparers and reviewers of the EIS by Section 101(2)c(iv): "The relationship between local short-term uses of man's environment and the maintenance and enhancement of long-term productivity." An examination of EISs will disclose that this topic is the least understood; it is usually the shortest and the least specific of all of the required sections. Here is a vital focus, however, of the EIS, even though the CEQ asserts that alternatives "are the heart of the environmental impact statement." The two prime alternatives are, indeed, those identified in 102(2)C(iv). To elaborate on this point, the Corps of Engineers suggest a three-part consideration (Jain et al. 1974):

a. Trade-off between short-term environmental gains at the expense of long-term losses . . .
b. Trade-off between long-term environmental gains at the expense of short-term losses . . .
c. Extent to which proposed action forecloses future options.

Section 102(2)c(iv) of NEPA might as well read "trade-offs between short-run economic interests and long-run environmental quality." Here there has been a substitution of "run" for "term." The word *run* has special meaning when used in the economic sense. In the long run, no factor of production is fixed or constant, implying the lack of any real constraints on resource-use decisions over a long period of time. Substituting the words "economy" for "uses" and "environmental quality" for "productivity" is logical, too. Thus, this subject, which must be addressed in every EIS, is the focus of concerns over often radically different viewpoints. To deal with the problem, one published assessment manual (Stover 1972) arbitrarily weights the future impact twice as heavily as any initial impact. Ultimately, the public decides.

But the level that the public viewpoint reflects is an important consideration; thus, it is on the local level that the difficulty is most evident. In a talk on federal pollution control activities entitled "Ecology and Economy," EPA Administrator Russell E. Train pointed out that environmental expenditures are no more inflationary than other types—that is, that all are more *dependent* upon the state of the economy generally than having inflation power of their own—and that expenditures for pollution control and other environmental quality control measures are, indeed, economical (Train 1974).

It is relatively easy for the federal government to spend a billion dollars knowing that the benefits to the public amount to more than that total. It is a great deal more difficult for the individual citizen to commit funds for benefits that may not accrue for a long period of time, and often not directly even then. A case in point is the threatened closing (by EPA) of a mill on grounds of noncompliance with air quality standards. The loss of tax revenue income to the federal government is probably undetectable; the loss of income from one of twenty mills owned by the company is certainly capable of being absorbed, or passed along to stockholders or product consumers who each have to bear only a small percentage of the total loss. But the loss of income to the individuals who work in the mill and to the local community is often untenable. The dilemma may be even more acute when considered in advance.

Is it possible, given the EIS that would accompany the government-ordered mill closing, to realistically discuss the trade-offs between short-term uses and long-term productivity? Not easily. Nor, of course, is there likely to be agreement on that portion of the EIS between the several reviewers. Is it any more possible, then, for a state-required EIS to realistically evaluate the trade-offs when a local contractor wishes to build a shopping center? Perhaps the evaluation can be achieved, but not all the EIS reviewers will be similarly inclined about this project either. In practice, such a decision is often reduced to a question for the local zoning board, which usually has unfortunate (and predictable) results:

> The public as a whole does not seem to be well served by current zoning practices, at least if we judge by urban sprawl, the depressing ribbon development along our major highways, and the pervasive ugliness of most American cities. Fundamentally, what is involved is the relationship between the individual and the community. It is my conviction that almost all current zoning ordinances in American are based on a faulty appraisal of this interrelationship, specifically between the economic rights and interests of the individual and the economic rights and interests of the community as a whole (Baer 1976).

What is more, individual zoning board members would probably like nothing more than to be able to serve out their term (and be re-elected) without a major controversy which, it can be almost guaranteed, will have two or more legitimate viewpoints, thus insuring everyone's unhappiness with any compromise settlement.

It is with zoning that professional planners meet frustration (Babcock 1966), resulting in planning either by variance or by default. Much of the controversial intercourse that takes place at public zoning hearings has, at its core, disagreements over the complex economic, environmental, and social impact of the proposed action. Here, too, confusion over the real debate abounds. If, for example, a local hearing on whether or not a gas station should be built on the corner of Fifth and Main gets bogged down in an argument over the policy decision of whether or not the town (or the nation) needs more gas stations, the EIS process is not working properly. An EIS concerned with whether broad policy should encourage more or fewer gas stations would discuss the trade-offs between some national policy-level environmental and long-range economic values that could be expected to be lost or gained. In contrast, concern over the site-specific action has more impact on the local social and biophysical environment. The more esoteric, national policy questions should not be a major part of the hearing or of the EIS concerned with locating a gas station at Fifth and Main.

Nor are controversies restricted to planning. Operational resource management law has been legitimately questioned as well:

> If the world is a space capsule, it seems rather silly to pin our environmental needs of the future to the robes of a legal system which stresses individual and corporate property rights, the exploitation of our resources, and the conquering of our environment.... [T]he laws needed to reduce demands on the environment cannot be obtained unless they are preceded by a more drastic change in attitudes.... If environmental quality is to be considered, it must be done at the outset. The costs and controls ... should be determined and assigned to the user, the government, and the general public.... In recent years, the public has become very critical and suspicious of resource agencies, especially the procedures and approach to the protection of esthetic, wildlife, and recreational interests. Yet the public must rely upon the fact that the agency, as charged by law, is protecting their general interests (Haik 1974).

But blind faith is not a viable solution; constructive participation is. And that participation must occur at all levels, in many instances with the EIS process as the vehicle of communication.

SUMMARY

On January 1, 1970, a new law culminated an evolutionary process of institutional adjustment within a nation that carved its existence and strength out of its resources. More or less concurrently, the sciences of resource development, protection, and enhancement kept pace with the demands of the population in its growth, wars, technology, and seemingly insatiable appetite for material goods and energy. The laws that controlled abuse of those resources lagged behind and often were out of phase with the sciences. When people discerned that the gap had widened too far, the National Environmental Policy Act was enacted. It called upon the creative, imaginative, and innovative talents of the population and challenged technologists to be environmentally sensitive and artistic—not in an unrealistic manner—but to apply their technology to the art of living within the spirit and letter of the law. It is ironic that the ecological crises we face have been the subject of literary giants often ignored by the "many pragmatically inclined students [who] were put off by the obscurely metaphysical, occultish notions surrounding the idea of harmony with nature" (Marx 1970). Now it is the write-up itself—the EIS—on which the law focuses.

Without attempting to paraphrase, the description of NEPA by Curlin and Hughes (1973) provides a succinct summary:

> NEPA is a self-executing, full-disclosure law which is analogous to the law governing security exchange transactions; it is intended to improve the planning process, expand public participation in governmental decisions and sensitize decision-makers to environmental considerations. The Act itself is broadly stated and subject to interpretation.... [T]he courts have made it clear that more than perfunctory or *pro forma* treatment is required, and that "purely mechanical compliance" with NEPA is not enough.... NEPA gave no authority to CEQ to approve or disapprove environmental impact statements. NEPA contains no explicit sanctions.

Further, like the U.S. Constitution, NEPA is vague and was quickly tested just as the former document is continually tested. During the initial years of its application, the federal agencies wished to see if they could get by without complying with the act, while the public flexed its newfound muscle. The result was extensive litigation, throughout which the courts supported, strengthened, and extended the act. Writing for Resources for the Future, Anderson (1973) states that

> from the beginning the courts have played a central role in enforcing NEPA's requirements.... The courts have required about three years to establish the basic trends for the first generation of NEPA issues.... In spite of some excesses in reading the Act too literally, NEPA's potential for lasting reform of federal government rests on the detailed judicial interpretations handed down in the first three years.... It appears likely that the courts will continue to bring NEPA to bear even more directly on the substance of agency decision-making.

The courts have.

Clearly, if NEPA has become an integral part of governmental planning and decision making at the project level, it will apply at the programmatic and policy levels, too, but not without support of Congress, the President, and the agencies themselves (Wichelman 1976). Generally, that support has been maintained. In the meantime, criticism of NEPA persists, but usually without complete understanding either of NEPA's underlying philosophy or of the nature of the EIS process. Nevertheless, NEPA's full potential is, in part, unfulfilled (Carter 1976).

Writing in response to a highly critical editorial in *Science* (Schindler 1976), former CEQ chairman Russell W. Peterson (1976) observed that

> as federal agencies adapt to EIS requirements, consideration of environmental impacts is becoming an integral part of decision-making rather than an afterthought. In addition, the EIS process opens up for effective review, both by the public and by government experts, decisions that were formerly made by individual agencies and their special-interest constituencies.

Stating it even more succinctly, Sullivan (1975) affirms that "NEPA and, in particular, the EIS, are here to stay." Thus, it behooves the environmental analyst to be aware of the laws so that the widely accepted goals of NEPA may be realized. Some of the laws provide specific criteria and standards by which to judge impact, while others provide sometimes simpler and more comprehensive means of achieving the spirit of the National Environmental Policy Act.

Compliance with the spirit of NEPA is, of course, a matter of attitude and faith. In order for the law to succeed, it is necessary to attain a positive attitude, which looks for improvement of the plan of action as a consequence of the EIS process and a firm belief in the fact that all facets of our life on this planet will be beneficially served by constructive implementation of the basic concepts that underlie NEPA's provisions.

CHAPTER TWO

Environmental Impact Analysis

Environmental impact analysis is the process by which environmental concerns are given consideration in the planning of any action. It is parallel to economic (impact) analysis which, in the case of water and related land-resources projects undertaken by or under the jurisdiction of governmental agencies, has come to be known as benefit-cost analysis.

As noted earlier, benefit-cost analysis evolved from the 1936 Omnibus Flood Control Act, whereas comprehensive environmental impact analysis is the direct result of a single piece of legislation, NEPA. Each type of analysis is a means of evaluating to what extent a proposed action is expected to contribute toward one of the dual objectives specified in the 1973 "Principles and Standards for Planning Water and Related Land Resources" by the Water Resources Council (38 Fed Reg 24779): national economic development (NED) or Quality of the Human Environment (EQ). To meet these objectives, environmental impact and benefit-cost analyses should be carried on throughout the entire planning process of an action.

There is another important parallel between the two tools of analysis: benefit-cost analysis must be, and usually is, conducted for actions that intentionally affect the economy, as well as for those that deal with natural resources, foreign policy, education, and so forth. Similarly, environmental impact analysis must be conducted for actions that primarily constitute environmental management and those seeking to achieve other purposes that inadvertantly affect the environment. Since environmental impact analysis came into existence as a result of major concern for the environment in the late 1960s and early 1970s, its focus has been on those projects and activities of government that have dealt with major environmental management actions. The authors of NEPA clearly intended and specifically stated, however, that *all* types of actions by the federal government would be subject to environmental impact analysis. Major inadvertant environmental intrusions (such as highways, canals, and airports) as well as intentional manipulations (dams, pesticide application programs, and stream channelization) have been the subject of much litigation and early environmental impact analysis. Thus, environmental impact analysis has been applied over a wide spectrum of actions, as is the case with benefit-cost analysis.

Much of the attention of the early years of environmental impact analysis focused on the physical, chemical and biological environment (hereafter simply refered to as the "natural" or "biophysical" environment), not the social-cultural environment, even though the latter was specifically identified for inclusion in environmental impact analysis by Section 102(2)A of NEPA. Now, after ten years of experience with the process, environmental impact analysis demands balanced consideration of the cultural as well as the natural environments. Note, however, that Section 1508.14 of the 1979 regulations specify that if only cultural impacts are anticipated an EIS is not necessarily required. Thus, emphasis remains on the natural environment.

THE INTERDISCIPLINARY TEAM

The single most important requisite for environmental impact analysis, specified in NEPA, court decisions, the CEQ regulations, and state legislation and guidelines is that it be conducted by an interdisciplinary (ID) team. The value of the ID team is reflected in the substance of the EIS, for each member contributes, in addition to technical knowledge, a bias. Political scientist David Abernathy's comments on clarifying biases can be applied to these technical biases, as well: "The trick is to make clear what our biases are so that we . . . can bounce them off against each other in ways that generate intellectual challenge and growth" (*Christian Science Monitor*, January 21, 1980).

The size of the ID team is not specified, nor should it be. To do so would be to obscure the fact that each environment is unique and therefore has different requirements for analysis of environmental factors. The composition of the ID team to conduct a given analysis demands primary consideration because its makeup may be far different if the project is environmental manipulation rather than an action that produces impact only incidentally. Thus, there has been a temptation to conduct environmental impact analysis with teams made up exclusively of environmental scientists for projects such as clearcutting in forests, creation of wilderness areas, or building nature trails. In fact, it would have been important to the success of such analyses to include cultural scientists on the team since the management decisions cannot be made in a cultural vacuum. By way of contrast, a team of environmental scientists (hydrologists, geologists, meteorologists, agronomists, biologists, etc.) might be quite appropriate for the determination of environmental impact for a project such as a shopping mall. It may be presumed that the sponsors of the mall have appropriately contracted for engineering services and for market and economic analyses in the initial determination of whether or not the mall should be built in a particular area and what its general specifications (size, anchor stores, configuration, etc.) should be.

ORGANIZATION FOR THE ANALYSIS

Environmental impact analysis consists of three phases, each one of which grades into the next to a greater or lesser extent depending, in part, on how the final determination is to be used and, in part, on the nature of the situation. The parts are organizing the job, performing the assessment, and writing the EIS.

The first phase is the definition of the problem and assembling the interdisciplinary team that is to conduct the analysis. The latter is dependent upon the former, as noted above. The definition of the problem consists of identifying the action and its scope, initial determination of relevant regulations and limitations on the part of all levels of government, and assessing public concerns. Management generally, or the project manager specifically, must also have some general feel for those portions of the biophysical and cultural environments that are likely to be affected and for which there consequently needs to be a specialist on the team. Without considerable experience in the process itself, the project manager alone is unlikely to be able to analyze the problems and pick appropriate specialists. Initial consultation with potential team members and relevant government and public officials is necessary.

A time frame for the conduct of the environmental impact analysis is necessary. It should be established and coordinated with the sponsor and potential team members, as well as any government

Job Name _____ ☐ EIS; ☐ EIA, LR ☐; Permit ☐: Wetland ▨
Contact _____ Organization _____ Indirect Source ▨
Address _____ Phone(s) _____ Impoundment ▨
_____ _____ Other ▨

Other/Alternate _____ IC # ▨

Job	Responsibility of (initials)	Chronology (enter dates)	Completed (✓) date
Contract			▨
Supplies			▨
Site Visit(s)			▨
Literature Review			▨
First reports due			▨
First Draft due			▨
Review with Client			▨
Draft EIS to Client to DEC			▨
Hearing(s)			▨
Final EIS			▨
other			▨

COST ESTIMATE

Team Member	Site Visit(s)	Subject Area/coverage/items/tasks	No. of man-days	Estimated Cost
Richards				$
Herrington				
Felleman				
Craul				
Black				
Manes				
				(rate)
		Subtotal		$
		Overhead, %		
		Extra Expense		
		TOTAL		$

Notes _____

34

agency that may be involved in permits, zoning board approvals, or environmental impact statement proceedings. Strict attention to relevant rules, regulations, and time requirements is essential to problem definition.

For a small analysis task, a single-page form (Figure 2) may be used to document particulars concerning the action and its sponsor, schedule the several activities that need to be accomplished (and provide for a check-off system to document completion of each job), identify participants and specify responsibilities, and estimate time and costs. The project manager should prepare the form and distribute copies to team members so that each knows what is going on, who is doing it, and when. For larger jobs, a longer form may be needed.

The brevity of this section belies its importance: indeed, it must effectively and efficiently set up the second and third phases.

ENVIRONMENTAL ASSESSMENT

The second phase is the actual environmental assessment or evaluation. Environmental assessment consists of the identification and evaluation of relevant environmental factors likely to be beneficially or adversely affected by the proposed action. Since this phase is the primary responsibility of the entire interdisciplinary team, communication between team members is essential. Successful completion of this phase results from a complex process that must be managed carefully in order to assure efficient and effective team operation, as well as integration of the various areas of expertise. A site visit is essential and should be done at least once with all members of the team on hand. Exceptions to this rule arise only if all members are acquainted with the site, or the action is not site specific.

A variety of techniques exist for the assessment, including those that are traditional in existing fields of expertise, those that are developing as a consequence of recognition of new environmental concerns, and newly available techniques for environmental monitoring. These are not treated here, for they usually are unique to a specific discipline, and all specialists on the team must be thor-

---

Figure 2 (facing page). Sample job sheet plan for a private firm. (Courtesy of Impact Consultants, Syracuse, NY.)

Figure 3 (over). Environmental impact assessment: flow chart and check list (FC&CL). (Courtesy of Impact Consultants, Syracuse, NY.)

I. THE ACTION _(describe in brief)_ _____

Sponsor _____ Date _____

STATUS a/

- ☐ Demolition
- ☐ Relocation
- ☐ Construction
- ☐ Operation
- ☐ Closure

TIMING

- ☐ One-shot
- ☐ Recurring

DYNAMICS

- ☐ Active
- ☐ Passive

PHASING

Periodicity: ☐ Continuously ☐ Hourly ☐ Daily ☐ Weekly ☐ Monthly ☐ Annually

Frequency: _____ number of _____ per _____

Time of day: _____ Hrs..., ☐ A.M. ☐ P.M. ☐ Both ☐ Irregular

Time of Year: ☐ Spring ☐ Summer ☐ Fall ☐ Winter ☐ All

TYPE OF SITE-SPECIFIC ACTION

☐ Project ☐ Practice ☐ Permit or License Required ☐

IS THERE CONTROVERSY ABOUT THE PROPOSED ACTION? ☐ Yes ☐ No.

APPROVAL(S) REQUIRED (Federal, State, Regional, Local)

Level _____ Specify Office/Official _____ Type _____

a/ Fill out FLOW CHART AND CHECK LIST for each entry checked. This is page _____ of _____ pages.

II. THE ENVIRONMENT _(describe in brief)_ _____

ENVIRONMENTAL FABRIC (Y = Yes; N = No; U = Uncertain; M = Can be Mitigated)

Atmosphere b/	Y	N	U	M	Y	N	Lithosphere b/	Y	N	U	M	Y	N	Hydrosphere b/	Y	N	U	M	Y	N	Energy b/	Y	N	U	M	Y	N
Weather	☐	☐	☐	☐	☐	☐	Soil:							Surface Waters:							Types:						
Air Quality*	☐	☐	☐	☐	☐	☐	Erosion/Move	☐	☐	☐	☐	☐	☐	Quantity	☐	☐	☐	☐	☐	☐	Coal	☐	☐	☐	☐	☐	☐
Particulates	☐	☐	☐	☐	☐	☐	Productivity	☐	☐	☐	☐	☐	☐	Quality†	☐	☐	☐	☐	☐	☐	Electric	☐	☐	☐	☐	☐	☐
Odors	☐	☐	☐	☐	☐	☐	Buffering	☐	☐	☐	☐	☐	☐	Regimen	☐	☐	☐	☐	☐	☐	Gas	☐	☐	☐	☐	☐	☐
Gases	☐	☐	☐	☐	☐	☐	Topography:							Ground Waters:							Geothermal	☐	☐	☐	☐	☐	☐
Noise	☐	☐	☐	☐	☐	☐	Slopes	☐	☐	☐	☐	☐	☐	Quality†	☐	☐	☐	☐	☐	☐	Oil	☐	☐	☐	☐	☐	☐
							Drainage	☐	☐	☐	☐	☐	☐	Quantity	☐	☐	☐	☐	☐	☐							

THE BIOSPHERE b/

	Y	N	U	M	Y	N
Flora	☐	☐	☐	☐	☐	☐
Fauna	☐	☐	☐	☐	☐	☐
Ecosystem	☐	☐	☐	☐	☐	☐
Endangered Species	☐	☐	☐	☐	☐	☐

b/ Adverse Impact Likely?
Are Original Data Needed?

Enter and rate THE FIVE MOST AFFECTED FACTORS from above and left.

Factor	(low) 1	2	3	4	5 (high)	Entry
	☐	☐	☐	☐	☐	
	☐	☐	☐	☐	☐	
	☐	☐	☐	☐	☐	
	☐	☐	☐	☐	☐	
	☐	☐	☐	☐	☐	

Average the five values and enter here and below: _____

DISCHARGE INTO:

	Y	N	Which?
Atmosphere*	☐	☐	
Landfill	☐	☐	
Water Body†	☐	☐	

Is an EXISTING STANDARD likely to be violated?

Yes _____ Describe _____
No
Absolute

CULTURAL ENVIRONMENT

Land Use Character	Commercial	Residential	Agricultural	Forest	Recreational	Idle
Site before action taken (B)	☐	☐	☐	☐	☐	☐
Site of finished Action (F)	☐	☐	☐	☐	☐	☐
Surroundings of Finished Action (S)	☐	☐	☐	☐	☐	☐

III. THE IMPACT

Type: ☐ Aesthetic   ☐ Non-Aesthetic   ☐ Destructive of Ecosystem(s)   ☐ Loss of Open Space

Classification: ☐ Neutral   ☐ Compensating   ☐ Additive   ☐ Synergistic   ☐ Catalytic

Evaluation

Classification: Rate from 1 (neutral) to 5 (catalytic)   (Low) 1 2 3 4 5 (High)   Entry
No. of horizontal spaces between S and F or B and F, whichever is greater.
Average of Five most Affected Factors.
Stability: ☐ stable   ☐ unstable   Rate instability
Chance of worst impact   (<1/10; 1/10; 1/4; 1/1; >2/3)
Sum of values at right _____ (sum)

Administrative Action   Date _____ Signature _____   compare _____ Remarks _____
Flow Chart & Check List
Negative Declaration Filed
EIA needed
Public Hearing needed
Full EIS needed

oughly acquainted with recent developments in their fields of expertise.

The prejudices and biases of individual disciplines cannot be dismissed lightly: they will in all likelihood persist and therefore should be recognized and discussed rather than ignored or eliminated (see page 70). Insofar as jargon is an inherent part of those prejudices, it helps to serve the final phase of environmental impact analysis if common language is used so that the public may ultimately comprehend the entire analysis procedure.

In the case of environmental impact analysis, the assessment consists of several subphases: first, there must be an initial determination of how severe any possible environmental impacts of a proposed action might be. This is normally done in the field, although it may be done in another location if the specific site is not relevant; but it must be done by the entire interdisciplinary team that is involved in the analysis procedure.

Assuming that the first subphase yields a positive conclusion, that is, that an entire EIS is to be prepared, the second subphase consists of the team's site visit, environmental characterization, and preliminary in-depth evaluation of environmental quality with and without the proposed action (Munn 1975).

Preliminary Evaluation

The team leader or project manager should guide the team through some sort of checklist, set of criteria, or relatively standardized procedure that will assure coverage of consequences of the proposed action that is as complete as possible. Attention must be focused, in particular, on the basic topics which are required for inclusion in an EIS by the Council on Environmental Quality's regulations (if the EIS is required by NEPA) or some other guidelines appropriate to the particular law being complied with, as well as agency regulations. Discussion of alternatives is especially important at this time and includes alternate uses of the site where the proposed action is to take place, as well as alternate means of obtaining the objectives of the proposed action and trade-offs between short-term uses and long-term productivity of the resources committed to the proposed action.

A flow chart and checklist (FL&CL) for environmental impact assessment (Figure 3) is proposed to assist the team leader or project manager in this important phase of environmental assessment. While some forms and checklists exist already, such as the New York State Department of Environmental Conservation's Environmental Assessment Form, some are lengthy, adapted to only one or a few types of actions, or are so all-inclusive as to be next to worthless for practical application in the field.

The form illustrated in Figure 3 had several criteria for its development. First, a single-page form was desirable; when completed, it would be easy to copy for each member of the team, would serve as the basis for a negative declaration or a finding of no significant impact (see Section 1508.13 of the 1979 CEQ regulations), and would be easily reproducible for dissemination to all potentially interested parties as a (legal) notice of the preliminary environmental assessment and determination of significance of impact. Second, it was important for the form to be generic in that it should not be tied to any particular type of project, such as construction. Third, it should touch base with all important characteristics of the action, the environment, and the impact; yet not be an encyclopedic checklist. It should also leave room for judgmental responses, identify where impact is expected, and call attention to those factors of environmental quality that would need field data or original on-site research for the completion of the assessment. Initial determination of possible mitigation of adverse impact needs to be included as well. Finally, where possible, the form should provide for some preliminary evaluation of environmental quality with and without the project and, as a consequence, some evaluation of the anticipated impact of the proposed action so as to determine what administrative action should take place.

The FC&CL is divided into four sections: the first three concern preliminary evaluation and determination of significance of potential environmental impact, the fourth section details any administrative action taken.

Section I, The Action, in addition to a brief description, sponsor identification, and date, treats seven general areas which assist in characterizing the action. *Status* refers to the type of action being undertaken. If more than one type is identified for any proposed action, a separate FC&CL should be filled out for each type; a new form should also be used to consider each alternative. The logic behind this is apparent in the subsequent consideration of *timing* and *phasing*, since an action that called for demolition, construction, and operation stages of some facility would each have different environmental impacts. Thus, the *timing* for demolition and construction would be "one-shot," while the operation would be a "recurring" action. *Dynamics* refers to a further characterization of the action which differentiates between a physical presence or activity and a "paper change," such as reclassifying land.

Phasing consists of the particularly important temporal characterization of the action: it involves periodicity, that is, the length of time between occurrences of the action; frequency, simply the number of times during that period the action is to take place; and the time of day and year. If the action is site specific, the form provides an opportunity to identify whether the action is a project or practice and if it requires a permit. There is just one question as to whether there is *controversy* about the proposed action: this is a threshold requirement for hearings under the CEQ regulations (Section 1506.6c). Under certain state permit procedures, a full EIS may be required along with a mandatory public hearing. Finally, the seventh part of this section on the action provides for identification of what governmental level(s) has *approval* authority, what office or officials are contacts or responsible, and what types of approval are involved. This information is important for reviewers of this preliminary assessment whether they are part of the interdisciplinary team doing the assessment, team supervisors or other involved government officials, legitimately interested parties, or the public at large.

Section II, The Environment, is divided into specific areas after a brief verbal description. These provide specific questions to answer concerning the *Environmental Fabric* (atmosphere, lithosphere, hydrosphere, and energy sphere) and the *Biosphere* (after Dee 1972). Two columns of Yes/No answer blocks are provided for each of the entries under the Environmental Fabric and the Biosphere sections: the first of these is to be used to answer the question, "Is this portion of the environment likely to be adversely affected?", and the second is to answer the question, "Are original data needed in order to conduct a complete EIS?" Columns for an uncertain response, and for an indication as to whether the impact can be mitigated are provided. Based on the collective judgment of the team conducting the assessment, the *Five Most Affected Factors* from the Biosphere and Environmental Fabric are to be listed and rated from 1 (low) to 5 (high) by checkmarks: the numerical value is placed in the column at the far right and the average of the five entered at the bottom of the listing of the five entries.

The fourth area is included to determine whether there are to be any *Discharges* into atmosphere, lithosphere, or hydrosphere, for such alone may trigger a permit, hearing, or full EIS. The same may be true if some *Standard* is likely to be violated.

The section on the *Cultural Environment* deals exclusively with land use character: as it currently exists (B), as the site of the finished action (F), and as surroundings of the finished action (S) are expected to appear. A checkmark in the appropriate column will identify each of these. If the finished action, for example, a park with a general rural characterization, was to be completed in an urban area, the number of spaces, measured in columns, between "residential" and "forest" would be the value to be entered in the Evaluation section below. If the horizontal distance (in number of spaces) between the condition of the existing and finished site is greater, that number should be used instead. This means of evaluating the extent either of land use change or of the continuity of land use integrates a great number of cultural factors that are subject to or are the cause of critical review in the evaluation of environmental impact, as defined in consideration of the land use spectrum (Table 1).

Five broad categories of land use are arranged from highly intensive to highly extensive use based upon human activity and natural resource conditions (variables). Derived originally from a simpler model involving land and water use for recreation (Clawson 1963), this model incorporates many of those attributes of the cultural environment that either are affected directly or are causative factors in impact. There is considerable oversimplification in this model; for example, diversity is not identified as being diversity *of* communities or diversity *within* communities. Further, diversity probably fluctuates, reaching higher levels between uniform land use categories. This increase in diversity as "edges" or in "transition zones" is of particular importance in consideration of ability to assimilate environmental disruption and, consequently, degree of environmental impact. A sixth category, Idle, is added to the FC&CL to reflect additional interactions and to provide five spaces.

While there may be extensive additional information required to completely assess cultural environmental impact (as in the case of the Environmental Fabric and the Biosphere), this technique highlights potential conflicts and should direct further study, if needed.

The land use spectrum model is not unlike the temporal one of Dansereau (1973), which is based upon the observable processes in natural and human alteration of ecosystems rather than on a checklist of characteristics. Thus, Dansereau's model is associated with trophic levels which relate energy and nutrient cycles. He identifies agents, processes, resources, and products, and develops the rank of human "escalation of power over the environment." Such a view is extremely useful in analyzing and assessing environmental quality, and the development of the model is available in several different publications by Dansereau (1970, 1971, 1973).

The prime value of the more spatially oriented land use spectrum in environmental impact analysis is either in perception of the action as a change in land use or as a means of evaluating the surroundings of the finished action: that is, the extent of change or the degree to which a land use is out of context with its environs. This can be quantified arbitrarily.

Section III of the FC&CL, The Impact, attempts to evaluate the

TABLE 1: Land Use Spectrum

	Class				
Variable	Commercial	Residential	Agricultural	Forest	Recreational
Characteristic					
Diversity	high →		→ lowest →		→ high
Natural systems	subjugated →				→ dominates
Resources	high human →				→ high natural
Population density	highest →				→ lowest
Management	user-oriented →				→ resource-oriented
Ownership	mixed →	mostly private →			→ mostly public
Purpose	mixed →	← single →	← multiple →		→ fewer
Economics					
Costs per acre	high →				→ low
Revenues per acre	high →				→ low
B/C value	short-term, visible →				→ long-term, ill-defined
GNP Contribution	highly productive →				→ low
Social					
Arts	most sophisticated →				
Communications	intense →				→ sporadic
Political power	high →				→ low
Government					
Commerce	intense →				→ sparse
Education	high →				→ low
Welfare	intense →				→ low
Services	intense →				→ low

impact in a variety of ways, some of which are capable of quantification and some of which are not. Some of the evaluations are already accomplished in Section II. Checkmarks in appropriate boxes are used to designate considerations about what type of impact, ranging from neutral to catalytic, is likely, implicitly acknowledging the fact that a great deal of clairvoyance is required to anticipate, much less predict, synergistic or catalytic impacts. The five categories are arranged somewhat arbitrarily, also given a number from 1 to 5, and the selected level is used in the Evaluation Section that follows.

In the summary *Evaluation*, checkmarks should be used (as before) in the appropriate columns, rating the impact from 1 (low) to 5 (high). This is already done to summarize 1) the classification, 2) land use, and 3) averaging of the five most affected factors, so the numbers merely need to be entered at the right. To these is added consideration of the stability of the existing environment and the chance of impact occurrence. The former should be either a collective "judgment call" by the team or an index, such as that presented by Dee (1972). Probability of impact occurrence should be entered for the most damaging impact and may also be the result either of judgment or actuarial data. This might be expressed in five levels of chance ranging from less than one in ten (1) to more than two out of three (5). Again, values indicated by the checkmarks are converted to numerical values in the extreme right-hand column of the FC&CL. These values are totaled (not averaged, as above). The total is then fit by comparison into one or between two of the five possible sums of values (which are arbitrarily set at the value that would result if all five lines were rated identically), and the minimum appropriate administrative action to take is suggested by an arrow leading into the last section of the FC&CL.

Whatever administrative action is undertaken as a result of this evaluation should be identified by date and signature. The filling out of the FC&CL is completed with the signature of a responsible or delegated official with his or her printed name, address, and telephone number. The number of copies of the completed form to be made is indicated, and to whom they are sent may be included on the reverse side. The FC&CL will be reasonably legible when reduced on a standard commercial copier on an 8½ x 11 inch sheet.

Whether this particular evaluation form is used or not, this subphase of environmental assessment must include discussion and documentation of those points just raised. Whatever device is used, it must permit viewing the environment in its totality as well as in detail and, most importantly, as a system.

The Environment as a System

Comprehension of the environment as a system is essential for successful implementation of environmental impact analysis, consisting of a broad, yet detailed environmental impact assessment (EIA) and the environmental impact statement (EIS) process (see pp. 13-15). The former is more than simple evaluation (or assessment) because one must evaluate that which is not there. That is, since the analysis must take place before the action that is expected to produce a possible impact occurs, the assessment must incorporate the identification of factors in the environment that are likely to be affected once the action is taken. Thus, EIA consists of the identification and evaluation of relevant environmental factors that are likely to be beneficially or adversely affected by the proposed action. The EIS reports that assessment. In order to be able to identify that which has not occurred, it is clear that the analyst must be able to envision how various parts of the environment interact, that is, how different parts are related to one another; this is the heart of systems concepts.

The system may be defined as "a functional whole, composed of organized, interacting, interdependent parts" (Russwurm and Sommerville 1974). The organization varies, its existence implying some sort of heirarchical structure, governed by natural or human laws. The interdependent parts may consist of elements or subsystems. The interactions that take place between those parts are processes: they may be physical, chemical, and/or biological. The character of systems, then, includes concepts of wholeness, function, parts, and dynamics. The complexity of organization and interactions and the number of parts and processes influence the stability of the system, a key concern in environmental impact analysis.

Properties of systems include boundaries, elements, interactions, and transformations of energy and matter. Once the boundary of a particular system is identified, the elements and subsystems may be identified. Experts can, upon identification of those elements and subsystems, further identify the processes (interactions) that exist in that environment and, consequently, the exchanges of energy and matter that are likely to take place. If the energy and matter circulate only within the boundaries, the system is said to be "closed," whereas if exchanges of energy and matter take place across the boundary, the system is an "open" one. All natural systems are open. Thus, inevitably, there are interactions between elements (and subsystems) beyond the boundary of the system being analyzed, and, consequently, identification of potential impact is made more complex and difficult to predict.

Since energy and matter move across the boundary of the system, the amount of energy and matter stored in the system is subject to change over time. It is the amount stored that is related to the stability of the system. Simply, a system with a large amount of energy or matter stored is resistant to sudden change by some outside force. That is, the system is not likely to be adversely affected by some action. This is because the large number of elements, with their attendant supply of energy and matter, can offer a wide variety of ways in which to absorb the intrusion or shocks of the action, buffering and modifying its effects. In other words, the impact is small. If one includes the variety of genetic material—referred to as the "gene pool"—as storage of large amounts of energy, it is clear that a biological system with great diversity of organisms will be more stable and less vulnerable to adverse impact than a uniform system.

Whether or not one is discussing biological elements in a system, the direct relationship between stability and diversity is fundamental to understanding environmental impact analysis. There is no single measure of diversity, but researchers in the different biological disciplines have created several indicators that may be useful in characterizing and ultimately predicting environmental stability and, therefore, impact.

It is possible to discuss the several parts of our environmental fabric (air, soil, water, and energy) separately and somewhat apart from the biosphere, insofar as they display varying degrees of diversity. For example, it is instructive to examine how each reacts to impacts of various types. Thus, one can commence the study of the specific sciences of interest by limiting the study to those aspects of the energy sphere, atmosphere, hydrosphere, lithosphere, and biosphere that are important from the standpoint of environmental impact analysis. Ultimately, the separate, isolated discussions are limited, for the spheres overlap considerably, and it eventually becomes necessary to stand back and look at the whole: the environment as a system.

Disciplinary Evaluation

The third subphase of environmental assessment consists of each individual team member's disciplinary evaluation of environmental quality with and without the proposed action. This subphase of environmental assessment is made up of research by the individual specialists in the field, the office, and the library, as needed. Individuals also should seek out other specialists, officials, and lay persons with specific knowledge of the site of the proposed action and its alternatives.

To acquaint individual disciplinary specialists with what they should be doing in their respective particular fields is beyond the scope of this volume. There are, however, some general principles of environmental assessment that may apply to evaluation of any environmental components in any given situation. Of particular importance are equipment, worst condition analysis, and aliasing.

Equipment

The equipment used in environmental assessment and EIS preparation consists of a wide variety of instruments, apparatus, and supplies that may be generally classified into "hard" and "soft" analytical tools.

Hard Analytical Tools. The hard tools include instruments of objective measurement used in assessing and monitoring environmental quality, specifically, the conditions of the environmental fabric, the biosphere, and pollutant levels and output. Under some conditions, certain of these instruments, such as surveying equipment, must be used only by licensed personnel. Standards for the care and use of instruments are generally available from the manufacturers, and in texts for professionals. These must be complied with since it is likely that data collected may be tested in court where strict rules of evidence may apply.

Selection of equipment should be based upon criteria of accuracy, precision, tolerance, and repeatability. *Accuracy* of measurement is the ability of the instrument to measure the true condition, whereas *precision* is an expression of the smallest unit of measure. *Tolerance* refers to the statistical probability distribution of observations made with the instrument, and *repeatability* to the capacity to reproduce those observations.

Certain types of equipment are well known and enjoy a firmly established reputation, but it is impossible to standardize equipment performance for all environmental impact analysis situations. Considerations of the environmental conditions and the purpose of measurements should dictate which instruments are to be used. In the case of complex chemical analyses of pollutants of the air and water, there is also the problem of where and when the analysis of the collected samples is to be done. This is important because some of the sampled materials may undergo significant real or artificial changes between the time of sampling and analysis if traditional standard methods are utilized in the laboratory, but varying degrees of accuracy and precision are sacrificed if field methods are utilized. The advantages of on-site analysis include the monitoring of readily identified sensitive sites and the opportunity to verify unexpected results.

Control over who uses the instruments is important to assure constant calibration and consistency of operation. Control also extends to methods of data collection, analysis, and reporting, so as to assure consistency associated with the maintenance of firm analytical standards. Again, the possibility of having to testify concerning an environmental measurement should underlie the care with which personnel operate and maintain equipment.

Soft Analytical Tools. The soft tools include checklists, matrices and other analytical software that aid the impact assessor by assuring complete coverage and maintaining perspective of the multitudinous environmental factors, actions, and impacts.

Checklists are available from several environmental handbooks (Corwin et al. 1975; Burchell and Listokin 1975) and government publications (Council on Environmental Quality 1972; Stover 1972; Environmental Protection Agency 1973, 1975). The matrixes discussed below also serve as checklists. The environmental impact assessment flow chart and checklist (Figure 3) combines preliminary assessment and a broad checklist.

Checklists have been arranged in a variety of ways for ease of use in the field and subsequent, more intensive analysis. One arrangement is that of a matrix, with actions listed across the top of a large sheet of paper, and environmental parameters listed down the side. This format is the basis of one of the first such matrices published (by Leopold et al. 1971) and provides a means of analyzing impact magnitude and importance. The suggested procedure conforms with CEQ's suggestion relating to "the probable impact of the proposed action on the environment . . . by providing a system for the analysis and numerical weighting of the probable impacts." The authors present various ways to summarize the magnitude and significance of impact for each interaction between a proposed action and environmental condition, thus providing, directly from the matrix, a collective assessment of primary interactions by a numerical indication of where the impact is likely to concentrate, spread out, produce compensating effects, and so on. Despite the fact that this represents only a limited portion of the impact's aspects, it is nonetheless a useful tool, particularly for new situations and/or personnel. It lacks criteria for assigning values (Munn 1975), however, and omits the probabilistic nature of both impact and secondary impacts.

A more complex matrix, involving secondary effects and feedback was developed by Sorenson (1972) and has been used in an intensive study of resource degradation and multiple use planning (Sorenson 1971). The format is keyed to uses (actions) that are associated with *causal factors* though that linkage may not be permanent. In fact, a change in method of use may effect a change in or an elimination (or introduction) of a causal factor that, in turn, will have some impact or change of condition. This type of impact is termed an *initial condition* and can be grouped or classified as biological, chemical, or physical; adverse or favorable; or by some other system. *Consequent conditions* are those that "describe the changes induced by the initial conditions that ultimately produce the *effect* or effects." The physical format of the matrix also allows for the display of *corrective actions* that are the "physical measures commonly employed to reduce or eliminate the adverse effects"; other measures, termed *control mechanisms*, including planning, zoning, regulations, etc., and references: "at least one specific reference to each *causal factor-condition-effect* relationship."

Another software technique is the use of overlays, developed into an advanced science by Ian McHarg (1969). Utilizing environmental qualities rather than assessing them directly, the McHarg technique involves the rating of certain critical environmental parameters from 1 to 5 and identifying the several categories on a series of overlays with progressively darker shades of color or crosshatching. By superimposing the overlays on a map, the path of least resistance for a right-of-way, for example, may be readily identified on the base map and transferred to the ground for more intensive on-site analysis. This and related techniques are discussed in Burchell and Listokin (1975).

The complexity of these and other matrix approaches make them extremely difficult to use without computer assistance and, even then, proper identification of inputs and interpretation of output is essential. The matrix is a useful tool in the impact evaluation process, but it is not a panacea. It provides the perspective needed to relegate the impact to its proper position with regard to the environment as a whole and to other impacts. Perhaps most importantly, it provides a framework for organization and, eventually, reporting the assessment. Used with discretion, it can provide the skeleton upon which to hang the meat of important decision-making information. It also can serve as a particularly useful guide to the beginning analyst, with the appropriate caution against overreliance on it as the sole means of environmental analysis.

Some of the available handbooks, manuals, or texts on environmental impact analysis are restricted in their application. For example, Yorke's dual matrix approach is designed for "planning and evaluating the impact of water development projects on fish and wildlife resources" and is used like a nomogram, directing the analyst's attention from action to physical/chemical characteristic of the water body, thence to "selected stream and floodplain biota" (1978). The handbook of Jain et al. (1974) is designed specifically for Army projects and activities and for use by Army personnel: it has little utility in other situations. Schaenman and Muller (1974) concentrate on cultural impacts of land development—with only minimal attention to air, water, and noise pollution—and on the amount and percent of change in "greenery and open space" and "wildlife and vegetation."

The approach of Hopkins et al. (1973) is intended for primary use in Illinois, but is written for and broadly applicable to the assessment coordinator. It advocates the use of "hierarchically structured lists" that suggest broad categories successively detailed at lower levels into specific considerations of the environment which may be affected. They provide both an organizational structure for the conduct of the analysis and a checklist for the team coordinator. Impact measurement in nominal (naming or identifying the impacted environment), ordinal (ranking of some environmental parameter), interval (unit change wrought by action), and ratio (unitless comparison) terms is also suggested; the first three are used in the flow chart and check list (see Figure 3).

Computer software, that is, analytical and data conversion programs, are tools that are useful to the analyst and are too numerous to detail here. Many are reviewed by Bennington et al. (1974). One broad-spectrum process, regional environmental management allocation process (REMAP), applies to either simple or complex situations in planning, locating, and allocating resources and in designing highways, electrical transmission lines, and other actions (Krauskopf and Bunde 1972).

Stellern et al. (1979) present a method to evaluate trade-offs between economic and environmental values identified in the Water Resources Council's Principles and Standards, and the complex problem of modeling multiobjective optimization is the focus of a recent report by Mades and Tauxe (1980). These and the results of further research on modeling and evaluation should also be helpful in assisting consideration of trade-offs between short-term uses and long-term productivity [NEPA, Section 102 (2) c (iv)].

The potential danger inherent in elaborate software systems is the user's overreliance on them. It is essential to know the premises upon which complex software systems are based; certain programs make use of basic assumptions about the local environment, that is, holding certain environmental factors constant while, in fact, they vary as a result of some action. The opportunity of utilizing computers, checklists, and other broad-spectrum analytical tools can lull the environmental analyst, the public, and decision makers in to the mistaken impression that environmental impact analysis can be a routine type of endeavor. Nothing is further from the truth. While it is possible to repeatedly build one of a series of hamburger restaurants in a chain because the basic design, engineering, and format for each are the same, such is not the case with environmental impact. Each environment is different, consists of different constituents in different temporal and spatial "mixes," and will react differently, depending upon local circumstances. Thus, there is a strong groundwork laid for the case against *pro forma* treatment of the environmental impact analysis and the preparation of the EIS.

Worst Condition Analysis

Worst condition analysis (WCA) can be defined as "interdisciplinary, on-site evaluation of a proposed action and its environment under that set of conditions which could produce maximum adverse impact." The word "interdisciplinary" is included because a team is more likely than an individual to succeed in identifying all possible bad situations and impacts; also, the team *must* be interdisciplinary by law. WCA is essential to constructive environmental impact analysis, especially to early identification and inclusion of mitigating and ameliorating measures and of design modifications to minimize impact. While WCA often carries a "doom and gloom" connotation and casts the proposed action in an unattractive light, it provides the proposed action with a clean bill of health if no adverse impact is disclosed. Finally, the word *could* is included in the definition because the analysis must consider the probability of occurrence of impact.

An important aspect of worst condition analysis involves how the action is to be used or operated. For example, dams planned for the Tioga and Cowenesque Rivers in northern Pennsylvania were designed for flood control and recreation. The floods of record in the area are caused by snowmelt, which occurs in March and April. At that time of the year, the reservoirs would be emptied so as to receive and hold back the floodwaters, to protect downstream communities. Had the dams been constructed and in normal operation, the reservoirs would have been full in preparation for the recreation season when hurricane Agnes occurred in June of 1972. As a consequence, the dams probably would have washed out with even worse flooding than did occur. But who expects hurricanes that far west? In June? Following a wet spring season?

Another example involves the proposed establishment of second homes around three small Eastern lakes with a drainage area of about 920 acres. Taking into account the precipitation falling directly on the lakes and net runoff from the watershed (analyzed with and without the proposed action, owing to construction of impervious roads, etc.), it was determined that annual runoff (2180 acre-feet) provides roughly five times the storage capacity of the lakes. Thus, it would appear that there is ample water to flush septic tanks and filter field pollutants out of the watershed. Careful analysis, however, indicates that only 11 percent of the annual runoff occurs during the four summer months when the second homes are in use and when as a consequence, most of the pollutants are released: thus, only 240 acre-feet of runoff are available during this critical period to flush a lake system that can store about 407 acre-feet. With insufficient water to flush the lakes during the warm, biologically productive summer months, adverse impact in the form of accelerated eutrophication is likely.

It may be advisable for the analyst to utilize several different levels of condition analysis, thus establishing a range of probable impact that may be more readily comprehended by the public. A complex example of such a study by the Environmental Protection Agency concerned the impact of vinyl chlorides on aquatic ecosystems (Hill et al. 1976). Whoever the decision makers are, they are better able to make wise decisions if they possess as much relevant information as possible about conditions under which impact may occur.

The major drawback to worst condition analysis is that, since it tends to amplify adverse impacts, it usually results in a negative-sounding report. The sponsors of an action are not likely to welcome derogatory observations about the impact of their project, and it really matters little whether the critical EIS is the work of an in-house department, a specialized agency team, an outside agency, or a consultant firm. Also, the public is more likely to recall low-probability high-risk events than more likely occurrences (Wilmot and Luna 1980). The direct consequence is likely to be concealment of information by the sponsor, antagonism, or an inadequate "rubber stamp" EIS after the fact. The best method to preclude antagonism between sponsor and critic is to assure participation by the EIS team in the early planning stages, with open lines of communication so that the EIS team gets all relevant information and the planners have the opportunity to modify the proposed action and include mitigative measures.

Aliasing

Aliasing is the consequence of monitoring cyclical environmental phenomenon at an interval that is constant and at least twice the frequency of the natural cycle. If the natural frequency is not known, sampling must be done on both a regular and irregular schedule to disclose it. While long-term measurements that are aliased may nevertheless identify a true mean value, the monitoring may miss maxima and minima, as well as the often more important periodicity of the environmental factor being monitored, thereby producing erroneous information and/or biased results (Munn 1975).

Where a parameter's distribution in time and space is not fully known, it is advisable to augment regularly spaced samples with intensive temporal and spatial sampling programs. For example, testing for phosphorous in a stream every Wednesday at 10 A.M. might not pick up the fact that a dairy cleans out its equipment and building every Friday. Nor might it identify another source of phosphorous from fertilized fields, the discharge point of which is downstream from the sampling point, even though the fields themselves are located upstream.

It is also possible to intentionally or inadvertently bias the data through aliasing. In one case, where the New York State Public Service Law required a year of weekly measurements, dissolved oxygen (DO) was measured "near the end of the day" every good-weather Tuesday throughout the summer months (DO is difficult to measure when it is raining and is of less significance when cold weather sets in). Regardless of the validity of the justification for the sampling program, it is clear that the results are heavily biased in favor of high DO: the prime source of DO is the photosynthesis of aquatic plants, an activity that peaks shortly after noon on sunny days. The sampling schedule guarantees the bias; aliasing may further contribute to assessment error. Not only does the absolute quantity of DO increase under such conditions but, with the slight warming of the surface waters and the consequent decrease in capability of that water to dissolve gases, the percent of saturation would also tend to be high in late afternoon. If such data are gathered for the purpose of evaluating the impact of introduction of organic wastes, thereby increasing the biochemical oxygen demand (BOD), or the impact of increased temperatures, the anticipated impact will be grossly underestimated. Only an intensive DO sampling program that ascertains the diurnal fluctuation on both overcast and sunny days will suffice to answer the rather obvious questions which arise.

An Example

This is a situation where environmental impact analysis would have identified adverse effects of the environment on the proposed project, not the other way around. But it illustrates a variety of important aspects of environmental analysis, including the benefits to be derived, the importance of the interdisciplinary team approach, communication of team members with each other and with project sponsors, and the value of having the opportunity to alter project plans during the planning stages. The Onondaga County Water Authority (OCWA) planned to (and eventually did) locate an intake at a depth of 40 feet in Lake Ontario, west of the City of Oswego, as shown in Figure 4. An aerial photograph clearly indicated that the prevailing westerly winds swept any Oswego River pollution to the northeast, away from the intake, because slight discoloration of the water marked the river water's location after entering the lake (a plume). The following facts would be disclosed and considered relevant during a thorough environmental analysis:

1. Winds do not always come from the west;
2. The conditions under which winds come from a direction that would lead the plume over the intake (easterlies) require a high pressure system to the north or a low pressure system to the south;
3. A low pressure system to the south is likely to be associated with storm conditions over the watershed of the Oswego River;
4. Storm conditions yielding rainfall would flush between-storm accumulation of pollutants from the stream system, causing a pulse of material to be discharged into the lake while the wind was out of the east;
5. The higher winds associated with the storm conditions could be expected to generate more turbulence than the prevailing westerlies, thus assuring delivery of pollutants to the 40-foot depth.

Expertise necessary to identify and evaluate these conditions would include a meteorologist, hydrologist, hydraulic engineer, and a limnologist. A check with the system designer or planner would be appropriate to provide additional information concerning phasing of the action: the system might not be used during the summer months, thus the entire concern would be irrelevant (the winter storms would be snow, and would not result in the flushing of the river system). And even the experts could be misled by the basic information, the aerial photograph: it could not have been taken when the adverse conditions exist; cloud cover associated with the storm and low pressure system would preclude a worst-condition view from an airplane.

Summary

The terminology of and relationships between the parts of environmental impact assessment are illustrated diagrammatically in Figure

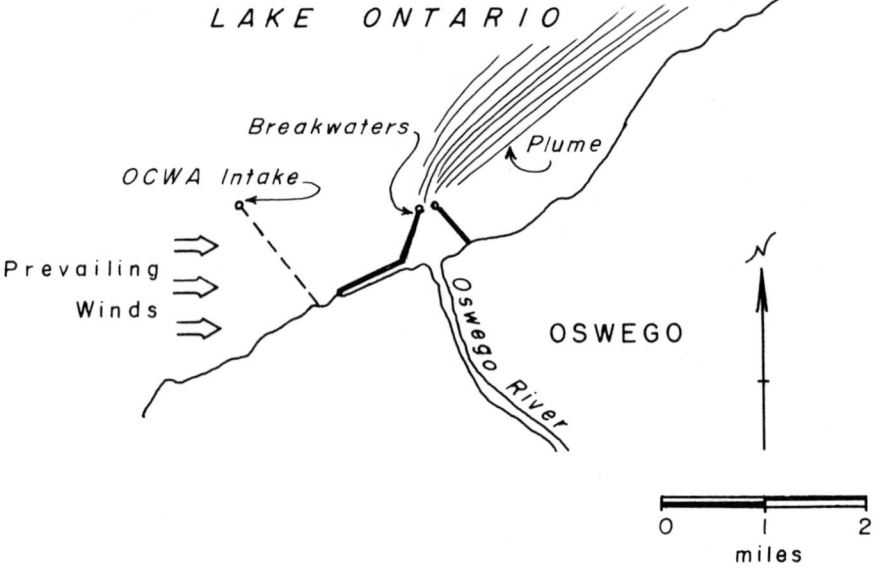

Figure 4. Oswego, New York and vicinity of Onondaga County Water Authority intake. (From National Oceanic and Atmospheric Administration chart 14803, 1978. P.E.B. 1/5/80.)

5. Environmental impact assessment is a gradually narrowing, spiral path that is initiated when a proposed action is considered in some environment. The three circles of environment, action, and impact are ringed with the various types of each. The overlapping areas between adjacent circles are labeled with the terms used to define environmental impact assessment (identification, prediction, and evaluation), each, notably, opposite the circle to which they primarily refer. Within each circle, special considerations, many of which appear on the Flow Chart and Check List (Figure 3), are listed. Others are discussed in this and other works on the topic and may depend, in fact, on conditions specific to the assessment at hand.

The conclusion of the assessment is the EIS. It serves as the focus of the communication process that takes place among all participants, in addition to presenting the results of the assessment and the record of decision, as required by law.

Figure 5 is at once oversimplified and descriptively accurate. Much detail remains to be added in order to complete the overall process of assessment, yet the compactness of the whole is evident. For programmatic EISs, the diagram is inadequate. Programmatic EIS require a continuing process of identification, prediction, and evaluation, not a terminating spiral. The elements, however, are in

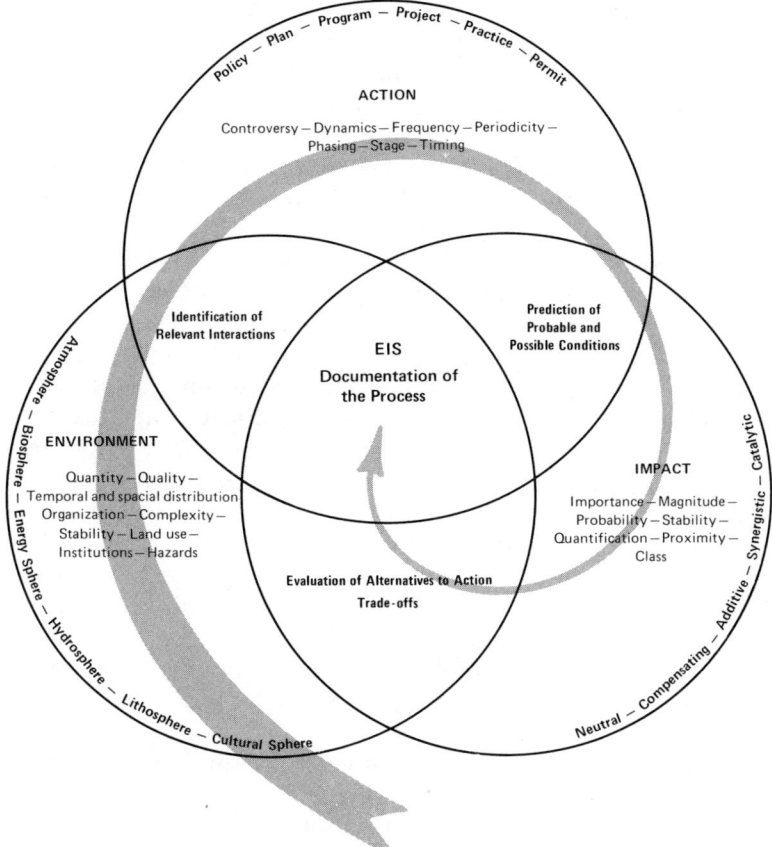

Figure 5. Environmental impact assessment: a diagrammatic representation.

their proper perspective, and all must be considered in order to constitute an environmental assessment. In sum, assessment essentially consists of asking and answering the appropriate questions.

PREPARATION OF THE ENVIRONMENTAL IMPACT STATEMENT

The third phase of environmental analysis consists of the public reporting of the results of the assessment. The laws and regulations that govern the environmental impact statement process are somewhat more specific on the characteristics which this phase must exhibit than on those phases that treat content. For example, the EIS must have a cover sheet with certain information thereon, should follow a specified format, and may be restricted in length. The EIS also must go through several phases between draft and final versions, including review by appropriate agencies, individuals, and specified waiting periods. Beyond these, there are no requirements, other than that the EIS must be written in a clear, concise, and comprehensible style and must be "analytic, not encyclopedic" (Section 1502.2, 1979 CEQ Regulations). The EIS reports the environmental assessment and, therefore, consideration of its content and format must begin during the assessment process, perhaps even on the first site visit by the interdisciplinary team.

When completed, the EIS ideally should reflect the fact that the assessment and statement were prepared by an *inter*disciplinary team and their concerted effort, not a *multi*disciplinary team and its separate and often disjointed contributions. Achieving the former is indeed an art, for it requires blending different approaches to the problem by specialists with a wide variety of prejudices, disciplinary backgrounds, and individual viewpoints, into a readable report that exhibits a high degree of integrity, perspective, and detail. It must flow; it must take the broad view; and it should not bog down in details so that one cannot see the forest for the trees. The third phase of environmental analysis may not always consist of EIS preparation: if the assessment phase indicates that no significant adverse environmental impact is anticipated, a negative declaration may be filed to document the findings of the assessment phase. Negative declarations are also subject to public review and must comply with agency and CEQ regulations for filing and reporting.

Criteria for an effective impact statement include that it must be deductive, concise, utilitarian, and documented. As its own example, the following consideration of form is presented in a style and format that fulfills these requirements.

Form

1.0 Environmental impact statements are most usefully constructed when they accurately reflect the environment being affected. To do this, they must build a special case that takes advantage of the natural ecology of the site, distribute the weight of the arguments in accordance with their importance in the environment and, finally, use an effective and documented style.

A Special Case

1.1 Take advantage of the natural ecology, unique to the site of the proposed action, to build a special case. This will include unique features of the ecosystem because a case so constructed can be defended on its own merits and generates a positive approach to the EIS process.

1.11 Unique features of the ecology include several of the subunits of an ecosystem: components, states of the components, relationships between the components, boundaries, and temporal and spatial proximities.
   a) Physical arrangement of the components means, for example, their presence or absence as a usual or unusual condition or a particularly unique combination of components.
   b) The state(s) of the component(s) in question needs to be examined critically as well. This is especially true if the state is variable, causing potentially different conditions to effect erroneous assessment or variable impact. For an example, see pp. 52-53.
   c) The nature of the relationships between ecosystem components also may be unique, as in the case of rapidly concentrating pollutants in a short food chain, a classic example of which is DDT residues (Wurster and Wingate 1968; DeLong et al. 1973).
   d) Boundaries deserve particular consideration here, for they are almost always unique, are usually defined as the limits of the involved ecosystems, and delineate the scope of the environment being assessed.
   e) The unique proximity of the action to any of the ecosystem characteristics (above) in time or space must also be a part of the description. Such considerations include the order in which the actions are taken, noting that the impact in all probability would not be the same if any two actions were reversed in time or location.

1.12 A case built upon its unique conditions can be defended on its own merits because precedents, although they should not be exclusively relied on, vary from case to case.
   a) Precedents for the particular type of impact may not exist at all, or may not be predictable in all cases.
   b) Precedents should not be relied upon, for the opposing view may find extenuating circumstances (that is, opponents may have a special case).
   c) Precedents dictate that the project sponsor will always take the same stand; this may, in fact, be undesirable, since considerations unique to the site of the action can legitimately require one position in one case and the

opposite or a different position under another set of conditions.

1.13 The building of a special case assures adoption of a positive approach to the assessment of the biophysical and cultural ecosystems, action, and impact, and should lead to reasonable allocation and use of resources based on informed decisions.

Representing the Environment

1.2 Distribute the weight of the arguments presented in accordance with their importance in the environment. This provides the detail to write an effective report and precludes reliance on a single issue.
    1.21 A well-balanced report is indicative of attention given to all relevant details.
        a) It illustrates first hand that the report does, in fact, represent the ecosystem (1.33a).
        b) The reasoned argument, in its proper perspective, carries its share of the weight and will, therefore, not attract an undue amount of attention, with attendant disadvantages (1.22).
        c) This concept should be applied as a general state of mind, in the interest of strengthening weaker arguments rather than "watering down" stronger ones.
    1.22 Single-issue arguments may be appropriate where the action is being opposed, and there seems to be little doubt that a basic objective or premise upon which the action is based can be successfully challenged, as determined by precedent-setting cases, changing conditions (including regulations and markets), or public attitudes.

Style and Format

1.3 Utilizing a comprehensible and logical style and format, prepare the statement in a deductive, documented fashion so that, if necessary, it will stand up in court to the rigorous requirements of examination and cross-examination of evidence. This will enable comprehension by a wide variety of readers, and simultaneously provide a utilitarian, deductive, and documented format.
    1.31 Prepare the statement in such a manner that it can be understood and used by a wide variety of readers.
        a) Anticipate vocabulary limits and capacities of such widely divergent readers as the lay public, experts in

your and other fields, superiors (who may have attained their position via a professional ladder different from yours), politicians, reporters, and lawyers (Rosen 1976).
b) Avoid jargon but, where necessary, use definitions, preferably included in the text at the first occurrence of the term; avoid using a glossary, a collection of terms limited to a special field of knowledge or usage, for it is too demanding upon the reader to study the meaning out of context in order to comprehend the text, much less put it to use.
c) Definitions should not confuse the reader, that is, do not utilize a term that needs further definition (Hayakawa 1974), nor use a term in its own definition.
d) In the process of explaining a technical problem, it is not necessary to start "in the beginning..."; present a reasonable discussion and then refer the reader who needs more background to an accepted standard, classical reference such as a text, which is used to bring students to a certain level of understanding.

1.32 This format lends itself to meeting this rather obvious requirement (1.31) and is suggested as a format for the detailed environmental impact statement to represent accurately both the generality and complexity of the environment yet be comprehensible by nontechnical readers.
a) Generally, the numbering system should not include more than about five subcategories of the next lower order number in order to facilitate overall comprehension. The fewer the better.

If one or more "added thoughts" need separating but are not one of several subpoints in the next lower tier of numbers, they may be separated as an un-numbered paragraph. The same may be done if a paragraph gets too long.
  i) It is preferable to leave excess blank space at the bottom of a page rather than have a paragraph "slop over" onto the next page; when this happens, the reader may have a difficult time identifying location or context.
  ii) Headings and subheadings may be used integrally with the format to set off principal ideas (1.3, for example).
  iii) Successive numbers can be clauses or elements of a sentence (1.32b), or completely separate sentences (1.32a).

iv) If the first paragraph on a page is not fully numbered, the next higher number should be shown at an index location.
b) The numbering system differs from that of typical engineering reports in two ways.
   i) There is only one decimal point, thereby reducing confusion.
   ii) The successive levels are indented to provide visual reinforcement of the numbering system and to facilitate reading.
c) The appearance of a large block of white space on the left does not signify wasted paper; a double-spaced, typewritten report takes about the same number of pages. Further, the large areas allow space for written comments and notes by the reader (Rosen 1976).
d) The nontechnical reader can get the sense of the arguments by reading the nontechnical expanded outline represented by the "hundredths-level" entries (eg., 1.32); the reporter might use the table of contents (units and tenths) to provide the perspective for a balanced story under the constraint of limited space; the politician, with a limited time for a large number of decisions, might only refer to the summary (prepared from the units-level entries which appears at the front); and the expert or concerned public critic might read the entire, documented text.
e) The generality or perspective of the ecosystem can be presented in simple terms, while the detailed description can be expounded upon in the text, with terms defined and references cited. Like the intricate interrelationships of the natural ecosystem, this format can be used to "spin a web" by inverse relationships between perspective and detail (1.2) and cross-referencing (1.4).
f) Finally, in this format, the EIS can be used to create a series of nested statements commencing with broad policy declarations and leading to specific actions on projects or practices (Black 1975). Such a "tier" system may be particularly useful in accommodating the large number of statements required for similar actions (Anderson 1973).

1.33 The requirements of formal logic and deduction lend themselves quite readily to inclusion in the format recommended.
   a) The statement must display the same organization,

ENVIRONMENTAL IMPACT ANALYSIS / 61

detail, and perspective as is observed in the environment being impacted (1.21a).
b) Categorical *propositions* of formal logic are identified by their *exclusivity*, *quantity*, and *quality*; these characteristics may be exploited for purposes of brevity, clarity, and precision, thus leading to definitive arguments relating to environment, action, and impact.
  i) All propositions (sentences or phrases) may be constructed in one of four types, which are situated at the corners of the square of opposition (Figure 6).
  ii) The exclusivity (precision and extent of the definition) of the term used (shown by closed or open circles), refers to identification of the object(s) in question (subject) and their relationship to all other objects (predicate).
  iii) Quantity relates to whether the proposition is about "all" or "some" of the items under consideration.
  iv) Quality refers to whether the statement is positive or negative.
c) The intricate relationships between various terms on the square are the subject of extensive study in courses in logic and the scientific method, or formal logic, and may be pursued further in Copi's *Introduction to Logic* (1969).
d) The requirements of an EIS admitted as evidence at trial are rigid. Thus, if the EIS is prepared to meet those standards (Sullivan 1975), it may preclude court action

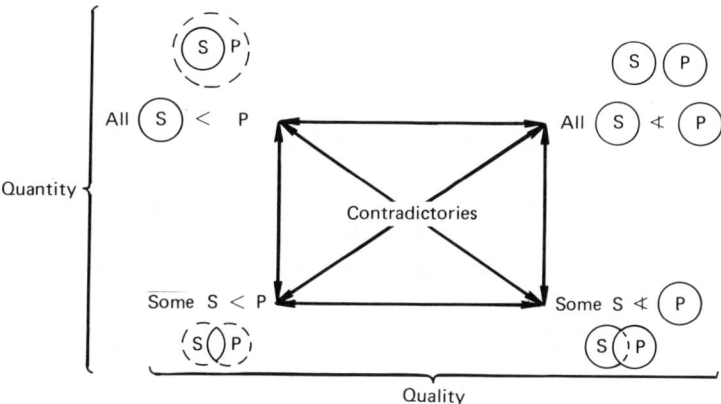

Figure 6. The square of opposition in formal logic. Symbol < means "is" or "are"; ≮ , "is not" or "are not."

itself—and the author(s) may avoid having to rewrite several versions for different readers. These requirements also mesh with those presented here as being desirable attributes of a competent EIS.

Documentation

1.4 Documentation, tracing any declaration back to some recognized authority, is essential; at the conclusion of each declaratory proposition, there should be documentation in the form of a reference to literature (see Section 1502.21 in the 1979 CEQ regulations), a cross-reference to another section (1.32e), or an original photograph, illustration, table, or graph that is, itself, fully documented. This meets the requirements of formal logic, as well, by presenting arguments in deductive fashion, that is, from the whole to the particular (1.33b).
  1.41 Documentation commences with selection of suitable maps which must be applicable in:
    a) Time and space, because the separation of a few miles or years may negate applicability, and
    b) Scale, which is essentially a problem of sufficient detail.
  1.42 Each original photograph, illustration, graph or table should be documented with name of preparer, date, location, scale, and source of data (as applicable) so that each is complete unto itself and can stand on its own. As an important example, consider the following recommendations on effective table preparation:
    a) If the table doesn't serve a purpose, forget it; it should be for the purpose of presenting field data, facilitating computations, illustrating results, presenting information concisely, or comparing observations;
    b) The table should follow the first reference to it.
    c) The table is "set" by the "stub" (left-hand) column heading, usually the independent variable or subject of the table.
    d) Balance the table physically to make it easy to read: group lines in 3s, 4s or 5s for appearance and ease of use.
    e) Document the table completely and show all units and sources of data.
    f) Be accurate; where no data were obtained use a dash not a zero, which is a precise value.
    g) Be consistent; line up decimals, use a similar layout for similar information, and, for values less than unity, show preceding zero in first row only.

h) Check arithmetic, as the reader who finds an error in the table (an event with a high degree of probability) isn't going to be too interested in or confident about the text.
i) Place all explanatory notes in table footnotes (identify notes with letters, so they are not confused with data entries), not in the text's numerical footnotes.
j) When in doubt, consult a reputable and relevant journal for style and format.

1.43 There is a wide variety of sources of information about the environment, including topography, soils, climate, hydrology, vegetation, and ecosystems. No single complete list of sources is available, but a rather thorough presentation is made in Burchell and Listokin (1975). Bibliographies are available at local libraries; state departments of conservation, fish and game, or natural resources; similar federal offices; high schools and colleges; and newspapers and professional journals.

1.44 Documentation includes that of the EIS itself, as required and specified by CEQ in the 1979 regulations (1502.11).

The Question of Size

2.0 The size of the EIS should be dictated by what must be presented in order to describe environment, action, and impact (Sec. 1502.15), not by a law that prescribes recording every known and potentially irrelevant parameter. Nevertheless, a page limitation has been provided in the 1979 Regulations (1502.7).

2.1 A format that serves as a checklist and scope of contents for the EIS is given in the 1979 Regulations (Sec. 1502.10), and a generalized outline that would comply with most nonfederal EIS requirements is presented on page 65.

2.2 For the preparation of an EIS under more detailed state or local laws, consult the appropriate guidelines for specific items that must be complied with in addition to those listed above. State laws and contacts are listed in CEQ's *Annual Report* (1979), in their six-year study of NEPA (Council on Environmental Quality 1976b), and in several of the handbooks and EPA publications (Burchell and Listoken 1975; Corwin et al. 1975; Environmental Protection Agency 1975).

2.3 Long before the 1979 revision of the guidelines, the Council on Environmental Quality felt that the size of the EIS needed attention. The observation still applies:

2.31 "In the future, it seems quite possible that the size of impact statements will eventually decrease.... Many impact statements now resemble encyclopedias. They discuss the project's setting in overly elaborate detail and contain lengthy descriptions of all species of plant and animal life in the affected area. Frequently, this reflects a lack of understanding of what is important and what is not. As the crucial environmental questions start to come into focus, it would become increasingly clear that much of this verbiage can be dispensed with" (Council on Environmental Quality 1975).

2.32 This prediction became reality as a consequence of President Carter's request for reducing paperwork and the subsequent revision and streamlining of earlier guidelines into the 1979 Regulations.

Substance

The five original topics to be included in the statement (according to NSPA) were the environmental impact of the proposed action, adverse effects if the action is undertaken, alternatives to the proposed action, trade-offs or relationships between local short-term uses and long-term productivity of the resources involved, and any irreversible and irretrievable commitments of resources caused by the action. A major omission was a description of the environment and the action, an oversight corrected by the Council on Environmental Quality in its original Guidelines for the Preparation of Environmental Impact Statements, issued in 1971. In addition, the guidelines indicated the EIS should include a discussion of the relationship of the proposed development to existing land use plans and potential offsetting federal actions. The former would refer to local and regional zoning, or to long-range plans for other government activities; the latter was to avoid charges that "the right hand doesn't know what the left hand is doing."

New York's State Environmental Quality Review Act (SEQR) added to the EIS presentation the discussion of potential impacts on energy and area growth, and the identification of ameliorating and mitigating measures that might be taken in order to minimize adverse impacts. With these additions, the EIS process became sufficiently cumbersome to warrant official attention to streamlining (see pp. 15-17).

The CEQ's 1979 regulations incorporate many important changes that do, indeed, streamline the EIS process and, on the whole, clarify and simplify the terminology and steps. It is evident that the process begun at the federal level has come full circle, for the new regulations include provisions that originated in state regulations and make innovative use of a variety of concepts found to be beneficial by government and private organizations at all levels.

For the general situation any EIS should include the following list of topics:

Cover page with EIS documentation, including list of preparers
Summary
Table of contents
   Description of the environment
   Description of the action
   Environmental impacts anticipated
   Unavoidable adverse impacts
   Alternatives to the action and uses of the site
   Trade-offs between short-term uses and long-term productivity
   Irreversible and irretrievable commitments of resources
   Interaction with existing land use plans
   Offsetting programs, practices, or plans of governments
   Impact on area growth
   Impact on energy consumption
   Mitigative and ameliorative measures
References
Record of decision, including list of EIS recipients
Comments, summary, and analysis of reviewers, if a FEIS
Appendixes

For each situation governed by law, the specific format for the EIS may be specified. The cover page should indicate the applicable law under which the EIS is being filed, along with all the required information indicated in the 1979 CEQ Regulations (Section 1502.11).

In order to apply the EIS process successfully, it is necessary to gather the essential elements as required by law and place them in proper perspective so as to prepare an EIS that will indeed, as NEPA requires," encourage productive and enjoyable harmony between man and his environment." In recognition of the widespread utility of and concern with the EIS process, an appropriate keyword is BROAD: brevity, relevance, organization, alternatives, and documentation.

Brevity

Aside from being mandated in the new CEQ regulations, the EIS must be concise for two commonsense reasons. First, the EIS must compete for decision makers' attention with simplistic expressions of economic, political, or national security considerations, or with other more complex emotional, aesthetic, and in some areas, archeological concerns. Second, the public is likely to rebel if it feels it is being "snowed" by a mammoth EIS, one that is too long to read and too technical to comprehend.

Relevance

The mandated brevity demands that the EIS deal only with relevant matters. Essentially, preparers of the EIS must ask, "Is this discussion (or point) helpful in decision making at the level involved?" or, "Is this information pertinent to the existing or anticipated conditions of the environment, action, or impact?" A variety of checklists, either separately or as part of complex analytical matrices, are available to assist in the preparation of an environmental assessment and the environmental impact statement that reports that assessment. Nevertheless, it is the responsibility of EIS preparers to determine which elements of the environment are pertinent, which elements of the action will interact therewith, and which impacts are likely to be significant. The challenge of determining relevance in the environment is essentially a problem of delineating boundaries—of ecosystems, for example—or of portions of the atmosphere, hydrosphere, energy sphere, or lithosphere that are affected.

Organization

In most cases, organization of the EIS must follow the new regulations' recommended format (Section 1502.10). Organization is an essential part of EIS preparation because it allows interested individuals to readily review an EIS. To that extent, the mandated format restricts the freedom and creativity of the preparers to allow each EIS to represent the individuality of the particular situation at hand. It is still possible, and usually desirable, to maintain such a balance of perspective and detail that the EIS accurately represents both the environment being affected and the environmental impact assessment itself. At any given point in the EIS, it must be self-evident where the reader is. Ideally, the EIS employs both "lumping" (the generalists' view) and "splitting" (the specialists' view) in such a manner that all possible important questions regarding perspective and detail are readily answered.

Alternatives

In the words of the new regulations, the section on alternatives (1502.14) "is the heart of the Environmental Impact Statement." The long list of topics to be included in the EIS, identified above, is separated by the new regulations into three sections, the first dealing with alternatives. The remainder are divided between the two subsequent sections. This highlights the importance of alternatives, which is what the CEQ intended.

Documentation

There are three forms of documentation that should appear in most EISs. The first is the documentation of the EIS itself, including who prepared it, for what agency, under what authority, and when. This information is mandated and appears on the title page or cover. The second type of documentation means reference to some credible source sufficient to verify or substantiate any fact presented in the EIS, including field data, photographs, tabulated or graphed information, or data sources from other locations. The third type of documentation is that which allows the lay reader to review a source of information that will enhance comprehension of basic areas of study and thereby enhance understanding the EIS. Thus, it is appropriate to include reference to classics in the disciplines to identify for the lay reader a source of information that allows better comprehension and use of the EIS.

Summary

The criteria established for preparation of an EIS that satisfies both the letter and spirit of NEPA are built upon the premise that the EIS is simultaneously the means for representing environmental vulnerabilities and presenting the environmental analysis for the review of all those concerned. Brevity, relevance, organization, alternatives, and documentation are identified as prime characteristics. An ability to put the facts and context into a report that is both comprehensible and readable is required. The real art is in providing the appropriate amount of facts on a correct background.

The balance of perspective and detail is the hallmark of the successful, that is, useful, environmental impact statement. A mere listing of environmental facts may be impressive, but without meaningful interpretation it is useless. The professional has an obligation to identify the important portions of the environment and to responsibly simplify, report, and document the analysis. If the professional cannot do this, the environmental analysis will be cast aside and decisions will be made on the basis of economic, political, social, or other considerations.

CHAPTER THREE

Public-Professional Relationships

The most expensive EIS and the assessment it reports is worthless if the EIS is not used. The EIS is intended to be used by the analytical team and by the project sponsors as a means of reporting the results of the analysis and, where appropriate, to illuminate mitigative and ameliorative environmental effects by incorporating modifications in the proposed action; by public officials to insure compliance with the law and to provide a record of the decision-making process; and by the public to be informed and to take an active part in planning and decision making. The EIS is a practical exercise in communication. This chapter considers the responsibilities and opportunities of groups between whom that communication takes place.

SPECIALISTS

The role of the specialist—the environmental scientist—in environmental impact analysis is threefold. First, the scientist must be the expert, knowledgable in a particular field or expertise, up to date on recent research and developments, holding credentials, and, if appropriate, licensed. Credentials, including degrees, experience, service, and publications, are also important when it is necessary for the expert to defend him or herself on the witness stand in formal proceedings or, as is quite often the case, before the informal review of a project proposal by a town board, conservation commission, or other group. On the stand, the lack of credentials poses a particularly sticky problem, for the only substitute is experience, and without the confidence that time on the witness stand affords, one often cannot get that experience. When interacting with a lawyer from the witness stand, it is a good idea for the expert to review the topics being revealed in advance so that there are no surprises and so that the lawyer may know what to expect in the way of answers and how the testimony might progress. Such a review should not be construed as an attempt to "put something over" on the public or the opposition, but rather as an opportunity to present information effectively and without confusion.

Second, the expert must function as a member of the interdisciplinary (ID) team. This entails constructive interaction with other experts that, in turn, requires some acquaintence with their fields of expertise and their jargon. Ideally, the necessity of explaining one's expert view of a problem to a fellow team member starts the communication process that ends with review by the public. This communication improves as team members work together.

It is important to differentiate between "multidisciplinary" and "interdisciplinary." The former implies that each participant "does his own thing," and the reports of each specialist are assembled and compiled, for example, by a project manager. The document that results from such a process is not an adequate EIS, for it neither integrates the expertise of the several specialists nor establishes communication between them; nor does it make beneficial use of interaction resulting from true communication between specialists. In contrast, the members of the ID team, in discussing various aspects of the environment, action, and impact *on the site together*, begin a process of communication that focuses on significant and relevant matters, and eliminates jargon so that the ultimate, integrated document will be capable of being understood by the public. Unlike jargon, which is usually unnecessary, the prejudices and biases that accompany an expert's education need to be understood by all; they cannot (and probably should not) be eliminated. On a practical level, the bias of each area of expertise must be recognized and productively exploited but only in concert with the biases of other experts. Perhaps the ideal state is one in which the team members assume each other's roles, an infrequent occurrence. A two-man team that accomplished this feat was able to identify the complex and elusive double helix structure of DNA (Watson 1968).

Third, the expert must interact with the public in several ways. During the course of an assessment it may be necessary to contact landowners, resource users, and interest groups that are legitimately concerned with the proposed action and/or the use of the site; good listening is a prerequisite here, for the public needs to feel that its views, complaints, or suggestions are included in the environmental assessment process (Heer and Hagerty 1977). The expert may also interact with the public indirectly through the media, a means of communication that requires considerable patience and tact. While it may be important to be candid and say, "I don't know," it is also important not to give the impression that the expert knows nothing or cannot express him or herself. Some discussion with other team members, project sponsor, and attorneys will avoid the interviewed expert's being trapped or disclosing information too early, that is, before its relevance, importance, and perspective are established. Finally, the expert may interact with the public at a hearing, on both a formal and informal basis. At certain hearings, members of the public may be allowed to question an expert witness directly: tact, good listening, and eye-to-eye contact are essential for an expert's successful response to such proceedings. The public may also attempt to contact members of the ID team during breaks in the proceedings for additional information, observation, or opinion. Witnesses should check with their attorneys regarding the extent to which comments may be made under these conditions. An off-hand, well-intentioned comment can ruin hours of hard work or formal examination on the witness stand, as well as the consulting firm's reputation, not to mention the client's position and financial investment.

As a team, the collective specialists can operate within the governmental agency that is proposing the action or as outside consultants. Neither NEPA nor the CEQ regulations stipulate where the team is to be located, although both the original version of NEPA and the amendment [Section 102(2)D] identify the "responsible federal official" implying the chief administrative officer of the agency. The amendment, as has been the interpretation of numerous court cases, asserts that regardless of where the EIS is performed, it cannot be merely "rubber stamped" by the agency head; agency review, project modification to include mitigative and ameliorative measures, and other inputs must be done by the project sponsor. Thus, the NEPA responsibility cannot be lightly waved aside or delegated.

The proximity of the EIA team to the project sponsor may take several forms, two extremes of which are shown in Figure 7. Figure 7A illustrates situations in which the team is subservient to the bulk of the agency; here it is often not in a position to either be convincing or effective in bringing about project modification and/or

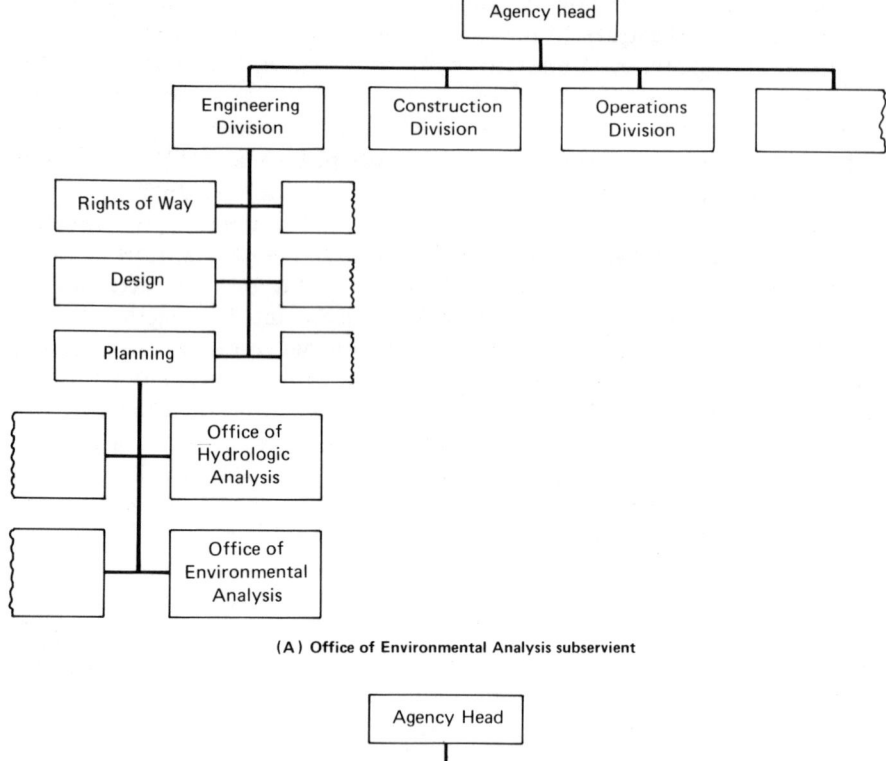

(A) Office of Environmental Analysis subservient

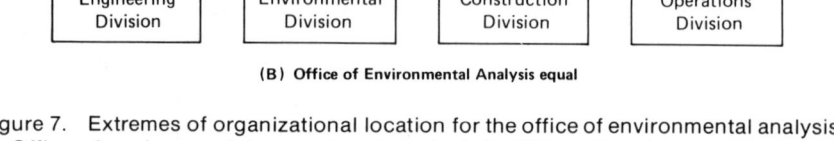

(B) Office of Environmental Analysis equal

Figure 7. Extremes of organizational location for the office of environmental analysis. **A.** Office of environmental analysis subservient. **B.** Office of environmental analysis equal at divisional level.

reduced adverse impact on environmental quality. At the same time, however, when it is so "buried" in the agency, the team is shielded from outside influence and may have better rapport with the design team. On the other hand, the team is exposed when it is on a par with the other primary units of the agency, as shown in Figure 7B, although it may wield more influence on the design and operations divisions. Maintaining perspective in either circumstance may be extremely difficult, but the implicit trade-off here is that the inside team may be better able to take part in the planning process than the outside consulting team. This is not always the case, for outside consultants can offer a degree of objectivity, flexibility, and creativity that is not possible from inside the agency.

A great number of EISs are now being prepared by consulting firms. The Corps of Engineers, for example, spends half of its EIS budget on outside consultants, EPA contracts nearly 100 percent of its EISs to consultants. Some of these firms have been in existence for some time (perhaps preceding NEPA), concentrating on one or another consulting service, such as architecture, economics, engineering, or forestry, and have added or gradually expanded into the EIS business. Some firms have responded solely to the opportunity that NEPA and state "little NEPAs" have provided and concentrate on a broad spectrum of environmental analysis consulting services.

The relationship of the members of the consulting team to the project's sponsors exhibits a few of the same problems cited above and introduces a whole set of new ones. First, it is apparent that if the consultants deliver a wholly negative and destructive EIS, they are not likely to be retained again by the sponsor. On the other hand, they cannot rubber stamp the client's plans because critical review will end either in rejection of the EIS or in litigation, and loss of reputation for the firm will result. Therefore, the consulting team must walk a tightrope between obstruction and whitewash (the National Association of Environmental Professionals refer to it as "opposition and advocacy"). Being outsiders, the consultants can often avoid the "insider perspective" (Henwood and Coop 1973) and thus be more objective than members of an in-house team.

A second inherent challenge is the ability of the consulting team to remain flexible as a group, adding or deleting areas of expertise depending upon the project at hand, yet to still draw on a broad and varied core of scientific disciplines. The team should also be capable of associating with other firms to form consortiums when particularly large proposals require a one-time, diverse review team. This flexibility presents problems of different sorts, such as logistics and travel, but, more importantly from the standpoint of EIA itself, making up a team as the job of the moment dictates may preclude experts' getting a chance to build their own confidence and ability to work with one another. The balance can swing the other way, too; a team can become too closed a group, impatient, unwilling, and even unable to communicate with outsiders.

A third problem relates to the team's attitude(s) toward the project and sponsor; although the EIA team should ideally come to the same conclusions about a proposed action regardless of how it approached a project or what its relationship to the project sponsor is, prejudice is likely to exist, just as a physician's diagnosis is influenced by what the patient first relates as symptoms. Minimization of effects of that prejudice can best be accomplished by adherence to scientific principles of on-site investigation and by review of relevant literature. In addition, the team and its individual members must be disinterested (not affected by the action or personal or vested interests) and be aware of their own latent biases. Identifying those biases and exploiting them objectively to put together a working team in which no one individual's bias dominates is a challenge to any professional group. The potential freshness and creativity that can flow from such an effort is a major benefit of a good consultant team's approach to environmental impact analysis.

Objective analysis of pertinent data and observations begins with the on-site inspection in a way that insures the balancing and offsetting of the biases of space, time, politics, and disciplinary viewpoint. First, the site must be approached from different angles and directions and by different team members, each with their own perspectives. Second, the site must be visited and viewed at different times of the day and year to ascertain temporally distributed conditions under which impact may occur. Third, the team should visit the site prior to consultation with project sponsors or opponents to formulate opinions without outside pressure or interference and to avoid the prejudices of others. Since this is often a practical impossibility, the team should consciously appraise its own bias as a consequence of reviewing plans with the project sponsor.

Reviewing a predraft version of the EIS (or permit application) with the sponsor is a controversial topic. On the negative side, the consultants must be careful not to compromise their evaluation of the situation or, even more importantly, give the impression (particularly to the public) that their expertise has been compromised. On the positive side, continuous communication with the sponsor of the action will awaken his or her environmental awareness and, in the long run, save money in lawsuits, changed plans, and adverse publicity. Most importantly, however, the EIS is supposed to be part of the planning process; thus, it is imperative that the sponsor's plans develop along with the environmental impact analysis. The same applies to continuous contact during the EIA with the public (Heer and Hagerty 1977). One of the beneficial results of this type of preliminary review is evident when some neighborhood, professional group, or clientele reacts to particular words as "red-flag" terms: these highlight an adverse aspect of an action and should be avoided because they may receive undue, disproportionate attention. The consultant may not be aware of such terms, and having a predraft review may serve to eliminate them without loss of professional standing. Professional integrity can be readily explained to the client if it is in fact, a characteristic of the team's members: it is not necessarily compromised by having the team accept input from the client, the public, or special interest groups or individuals. This topic is the subject of a major portion of the National Association of Environmental Professionals' Code of Ethics.*

The specialist—occasionally a generalist, editor, or project manager—who coordinates the analysis and guides the EIS in its creation and review must have great capacity to listen, an ability to synthesize the EIS, and a willingness to stick his or her neck out. The first requirement embraces the team session at which each member is assigned (or agrees to take on) certain portions of the EIA and EIS preparation. Subsequently, each of their drafts must be blended together and, as appropriate, cross-referenced. The in-house draft must exhibit coherence, integrity, and the logical organization that is characteristic of a single individual's effort, even though it is basically a collection of several separate papers or segments.

The first draft must be circulated to team members to allow clarification, correction, and comment on any details as well as feedback on overall perspective, tone, and conclusions. Often, the comments are highly critical and the coordinator must be able to accept that criticism gracefully and not take it personally and then act on it constructively by redrafting. This can be an intricate job, because cross-referencing must be kept in mind, often over extended periods of time. For example, in the first draft the team ecologist may have commented on the agronomist's discussion of soil fertility. A subsequent adjustment of the soils discussion, suggested by another team member, should trigger another review by the ecologist who will not necessarily look at the particular section again unless the coordinator calls it to his or her attention.

Lastly, the coordinator should then be able to testify, at a hearing or court trial, as to who is prepared to testify on what subjects, as well as coordinate their testimony with the attorney. The coordinator must work closely with the team's or the sponsor's attorney(s) throughout the hearing. Here is where the traditional qualifying and/or thesis defense exam, which is a part of the advanced degree, is vital experience. Extensive and detailed questioning, hostile cross-examination, and sometimes vicious attacks on one's professional capabilities must be borne with professionalism; candor, sincerity, and a sense of humor are necessary attributes, not to mention a high degree of self-confidence without arrogance.

---

*The Environmental Professional 1(1):ii. National Association of Environmental Professionals, P.O. Box 1223, Alexandria, Va. 22313.

The sequence of operations by the coordinator and other actors in a hypothetical but typical EIA-EIS proceeding is shown in Figure 8. Upon agreeing to perform the EIA, the coordinator, attorney, and sponsor should draw up a contract that identifies services to be provided, fees to be paid, and deadlines to be met, as well as the personnel involved and the relationship between client and consultant. It is likely that the process is unique in its details for a particular proposed action and certainly for different government levels and/or legal compliance situations. The nature and complexity of the job, however, is apparent.

OFFICIALS

The public officials who are entrusted with the maintenance or improvement of environmental quality have a difficult task. On the one hand, they are usually in such a position because of a high degree of motivation and dedication to high ideals, for their remuneration and the public's opinion of them is not high. On the other hand, as is the case in several states, there may be a great degree of power in the hands of bureaucrats, and that power tends to be a corrupting influence. Corruption can take several forms, from perpetuating the agency's activities to illegal contracts, kickbacks, and conflicts of interest. Even if there is no corruption, if the public thinks it is there, the bureaucrats lose their effectiveness.

Further, the government conservation official responsible for evaluation and/or protection of a particular piece of public real estate, a wetland, for example, may become so enamored of it that a landowner-land relationship takes over and clouds the judgment and perspective of the official and that of his or her superiors. Such overzealous protection of public property in fact acts on the behalf of one public at the expense of another. Thus, the official becomes partisan, and governmental prestige—confidence in the agent or agency—as well as the original environmental goals are jeopardized.

At the federal level, agencies have a wide variety of functions. Table 2 lists 21 agencies selected because they are directly involved in environmental concerns. Some of these agencies with a wide range of functions consequently have considerable power, for example, EPA, TVA, and the Forest Service. Similar situations exist at the local and state levels, with a high degree of variation from state to state. The public, the specialist, and the official must all be aware of the extent of opportunities, restrictions, and challenges that emanate

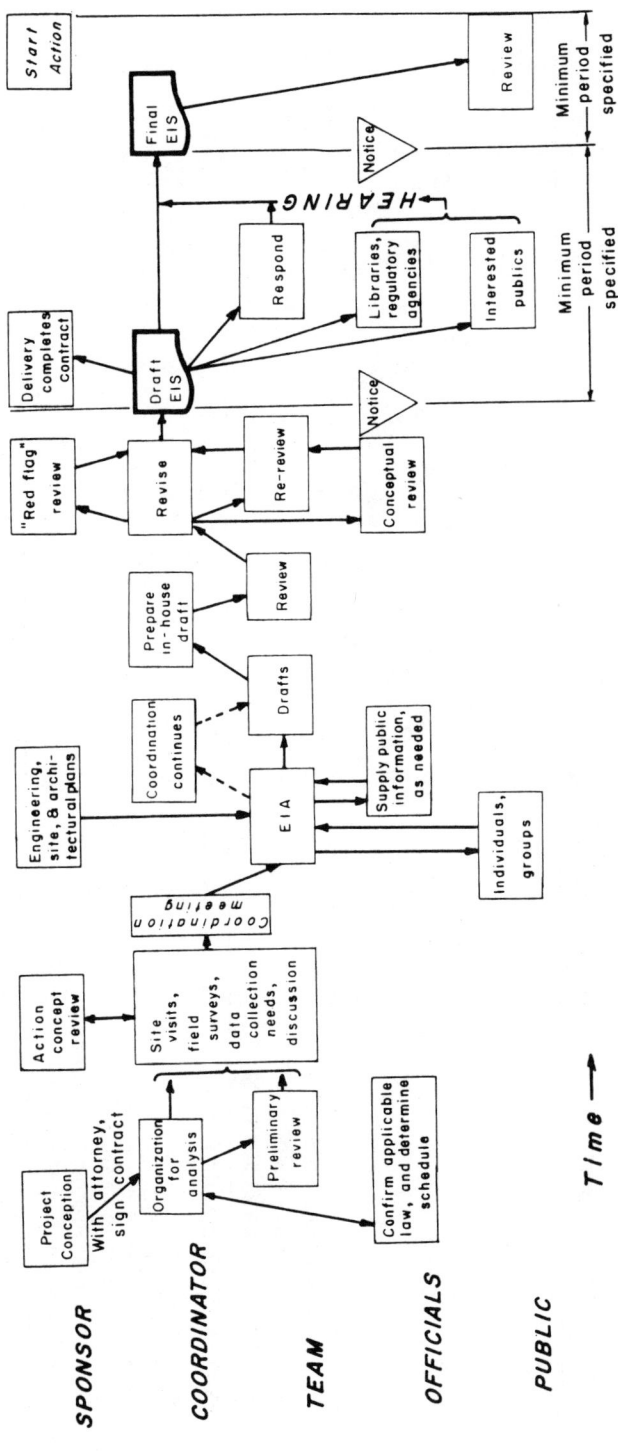

Figure 8. Typical flow of activities in the preparation of an EIS, focusing on a consultant team coordinator.

TABLE 2: Primary Functions of Selected Agencies

Agency	Land administration	Coordination	Enforcement	Operations	Planning	Policy & promotion	Regulation	Research	Service	Study group
Agricultural Research Service								✓	✓	
Bureau of Indian Affairs	✓								✓	
Bureau of Land Management	✓								✓	
Bureau of Reclamation				✓				✓	✓	
Coast Guard		✓	✓	✓			✓	✓	✓	
Corps of Engineers		✓	✓	✓	✓	✓	✓	✓	✓	
Council of Environmental Quality		✓				✓				
Environmental Protection Agency		✓	✓		✓	✓	✓	✓	✓	
Federal Emergency Management Agency		✓				✓				
Federal Energy Regulatory Commission						✓	✓			
Federal Interagency River Basin Committee*										✓
Fish and Wildlife Service	✓							✓	✓	

TABLE 2: Primary Functions of Selected Agencies (Continued)

	Land administration	Coordination	Enforcement	Operations	Planning	Policy & promotion	Regulation	Research	Service	Study group
Forest Service	✓	✓				✓		✓	✓	
National Atmospheric and Oceanic Administration		✓						✓	✓	
National Park Service	✓									
Nuclear Regulatory Commission			✓				✓	✓		
National Water Commission*										✓
Public Health Service						✓		✓	✓	
Soil Conservation Service				✓	✓	✓		✓	✓	
Tennessee Valley Authority	✓		✓	✓	✓	✓		✓	✓	
Water Resources Council		✓			✓					

Source: National Archives and Records Service. *United States Government Organization Manual, 1971-72*. Washington: Government Printing Office, 1972 [subsequent volumes entitled *United States Government Manual*].
*Agency no longer in existence.

from governmental structure and mandate. Those seeking to work as professionals with officials or those seeking to "watchdog" government should invest in the appropriate governmental manual that typically lists agencies, personnel, office addresses and telephones, general responsibilities and jurisdictions, and charter legislation for more detailed reference.

Whether the EIS is prepared by agency or consultant, there is the possibility of *pro forma* treatment. Thus, one of the greatest dangers inherent in NEPA is the possibility that individuals and bureaucracies, wanting to insure their futures or to pacify the public, will create "standardized" EISs for a particular type of project or program. There is a great temptation, once a successful EIS has been prepared, reviewed, revised, and defended, to dust it off when the need arises for an EIS on a similar project. In view of the fact that there are no specific guidelines, rules, or laws preventing such misuse of the EIS process, it can happen and probably already has. In order for a government agency to get away with such behavior, however, there is one essential ingredient: a pacified public. If the public maintains interest and an active role in the EIS process, the process will not be so abused.

But NEPA demands more than *pro forma* treatment, and the act is having a substantial positive effect on federal agencies. Wandesforde-Smith (1971) summarizes an informative article by stating that

> most resource management agencies realize that they are in deep trouble ... essentially political. ... Since few bureaucracies readily accept a reduction of their budget or a drastic reorganization of their programs, it is not surprising that existing agencies are looking for ways to ensure their survival, and perhaps to improve their position by putting essentially unchanged programs in a new linguistic wrapping.

Wandesforde-Smith points out that NEPA has played a major role in bringing this situation about and doubts that simple lip service and minor adjustments will suffice. He gives some indications that "the politics of the environment" are really different from "the politics of resource management" and adds that if they "are fundamentally different in kind ... then a theory of environmental administration will be required that both reinterprets present experience and provides the basis for some drastic, long-run alterations in administrative practices."

CITIZENS

It is clear that the creation, passage, implementation, and fulfillment of NEPA has in large part resulted from citizen concern and involvement. In order to maintain a safe and healthy environment and simultaneously to maintain and upgrade the standard of living for our population, it is also clear that the public will have to continue to be responsibly involved. The EIS process provides that opportunity.

On the bureaucratic side, "the federal government has taken the lead in broadening the basis of public involvement in resources and environmental policy-making" (Sewell and O'Riordan 1976). According to Wengert (1976),

> There is no substitute for a policy which seeks the public interest.
>
> For some time after World War II it was fashionable among social scientists to assert that the public interest was a myth—like religion, an opiate of the masses. What was confused in this view were the differences in defining the public interest, with the much more important fact that it was the *search* for the public interest, the *requirement* to *rationalize* decisions as being in the public interest, that was the significant aspect of the concept. The preacher says "Seek ye first the kingdom of God"; the responsible democrat says "Seek ye first the public interest." Neither is easy; with respect to both it is the *seeking* that makes the difference, even when it is recognized that we often fall short [emphasis in original].

It is important for the individuals within the bureacracies to understand the role of the citizen and how public interest and public opinion differ. The public interest is what is "good" for some defined group of citizens as perceived by an observer, whereas public opinion comprises "the opinions of the aggregate of individuals making up the public under discussion . . . specific to a particular set of conditions" (Truman 1959). Within this context, citizen participation can be for a variety of purposes (Wengert 1976)—as a "sound and desirable" policy inherent in good government and a democracy; as a strategy "to accomplish . . . objectives"; as a means of obtaining valuable inputs to the decision-making process by communication; as a means of resolving conflicts; and as a means of therapy. Any or all of these may be important in any particular situation, and the effective government official involved in environmental decision making must be sensitive to his or her publics' needs and wants.

On the public's side, individuals must provide the appropriate reaction to the agencies' activities and respond in a responsible manner. It is essential that publics keep informed politically, to understand the pressures placed upon the decision makers in the agencies and the legislatures; ecologically, to appreciate our natural environment and heritage; and technologically, to take advantage of new developments and broader comprehensions of our way of life.

Generally, the public must respond to agency proposals with an understanding of legislative charges, pressures, and agency responsibilities and opportunities; it must participate in the EIS process in a constructive manner, that is, seeking solutions to problems, not creating them; and it must adopt a practical, realistic approach to conservation, a balance of exploitative and preservation activities that match use over time with supply over time (Black 1969)—so that we and future generations may, indeed, live "in productive harmony."

Citizen participation, then, is the key to the success of the environmental protection legislation such as the National Environmental Policy Act and all those modeled after it (Orloff 1978). In order to live up to this charge, individuals must exhaust all administrative channels in seeking a solution to a conflict prior to gaining standing in court, if that becomes necessary; must keep informed and participate in hearings, where available; and must not be obstructionist, that is, objecting to proposed actions solely for the purpose of showing or gaining power. More specifically, effective citizen action consists of defining the problem, becoming informed, developing a plan of action, gaining support, and taking constructive action to achieve certain specified objectives (Stapp 1975).

There are benefits that accrue to the public and community at large from environmental impact analysis. As identified by Henwood and Coop (1973), these include more informed decision making; better integration of public values and interests into the planning process; more fully documented administrative decisions; periodic reevaluation of basic planning assumptions; and considerable "spin-off" from applied research into impact analysis and environmental monitoring.

HEARINGS

Although it is primarily a device to force the agency to examine the anticipated impact of its own action, NEPA also forces public review and, as a consequence, provides an additional role for the public hearing.

The public hearing provides the interested, concerned, or disgruntled individual with the opportunity to be heard and allows the citizen to hear the more detailed plan of action by government. While it is sometimes difficult to ascertain whether or not someone is listening, the value of the hearing remains. It offers a challenge to present views and information in such a manner that someone will listen. Although not specified in NEPA, it is usually the most expeditious means of obtaining "public comment" and was specifically identified in the guidelines of the Council on Environmental Quality (1972). Hearings are implied to be a normal means of receiving comments in Section 1503 of the current regulations.

Derived from the town meetings of colonial days, "hearings are a viable and useful part of the democratic process at every level of government" (League of Woman Voters 1972). Further, the hearing can be the focal point in the communication about the EIS for a proposed action. Although some feel otherwise (for example, White 1972b), Ross (1972) maintains that "the hearing itself is a matter of great importance to all parties, and is probably the most vital protection the public has." He goes on to point out that "the public is being asked to accept on faith the soundness of far-reaching decisions involving complex technological, scientific, and social factors. No one is going to agree to make the necessary sacrifices without confidence and trust." There is responsibility on both sides when it comes to the public hearing; every participant must be willing to hear as well as to speak. With its emphasis on the interdisciplinary approach, NEPA places new challenges before this much-abused process.

The purpose of the hearing will dictate format and opportunity of participation. The pamphlet "Anatomy of a Hearing" (League of Women Voters 1972) identifies six general purposes, which can be arranged into three types: to receive, to relay, and to review information. In the first group are hearings held solely for the purpose of general communication, "to find out what citizens think, to get expert analyses and data, to highlight the issue for better public understanding," and so forth. Government officials may wish to hold a hearing to get feedback on a specific issue, action, operation, or practice. Hearings may also be held for the purpose of solving a specific problem, to get information which may be used in drafting legislation, for example.

The second group involves government officialdom giving out information. At best, such a hearing may be held for the specific purpose of providing the citizenry with the information necessary to choose between alternatives in a referendum or of paving the way for a later hearing under the first group category. At worst, such a hearing is merely a rubber stamp process for an previously implemented government action.

The third group has hearings of only one purpose: to examine alternatives. The alternatives may be different ways of resolving a problem, the solution to which is mandated at some higher level while the details and methods are left to local concerns, or perhaps whether or not to take a specific action at all. Either case fits the review of EISs and the details of an environmental impact assessment upon which it is based.

The league's pamphlet contains many examples of hearing settings, some do's and don'ts for participants, and some introductory ideas on how to go about speaking and listening at the hearing. If the purpose of the hearing is kept in mind while preparing testimony, the hearing can indeed be a useful procedure.

CONCLUSION

Two key words in making effective use of the environmental impact statement are "constructive" and "responsible." Their relationship to each other is ramified throughout a quote from an unpublished speech by Regional Forester Charles A. Connaughton in 1963 when he pointed out that the public and the professional need each other:

> The professional, trained and knowledgeable in terms of technical adequacy, needs the influence of the public to be sure that the resource use and attitudes are in tune with general opinion. If they aren't, no matter how technically sound the decision, it is doomed to failure. On the other hand, public attitudes which do not take the professional recommendation into full consideration are liable to be unattainable and doomed to frustration.

The public hearing doesn't prevent this potential failure in communications: it only provides the opportunity to do so. As such, the hearing is a vital and essential part of the EIS process. And the EIS process provides a vehicle for effective communication between all participants.

Bibliography

Anderson, F. R. 1973. *NEPA in the Courts*. Baltimore: Johns Hopkins Press.
Babcock, R. F. 1966. *The Zoning Game*. Madison, Wisc.: The University of Wisconsin Press.
Baer, R. A. 1976. "A New Approach to Development Rights and Zoning." Conservation Circular 14:109.
Baldwin, M. F. 1981. Statement Before the Subcommittee on Fisheries and Wildlife Conservation and the Environment, Committee on Merchant Marine and Fisheries. February 25 (Mimeographed).
Barbaro, R. and F. L. Cross, Jr. 1973. *Primer on Environmental Impact Statements*. Westport, Ct.: Technomic Publishing.
Bennington, G., S. Lubore, and J. Pfeffer. 1974. *Resource and Land Investigations (RALI) Program: Methodologies for Environmental Analysis*. Environmental Assessment, vol. 1. NITS PB-244600. Washington, D.C.: Mitre Corp.
Black, P. E. 1969. "Confusion on Terms." *Water Resources Bulletin* 5(4):8.
Black, P. E. 1975. "Environmental Impact Statements in Planning Water and Related Land Resources." *Water Resources Bulletin* 11(5):881-886.
Brown, H. 1954. *The Challenge of Man's Future*. New York: Viking Press.
Burchell, R. W. and D. Listokin. 1975. *The Environmental Impact Handbook*. Center for Urban Policy Research, Rutgers University, New Brunswick, New Jersey.
Carson, Rachel. 1962. *Silent Spring*. Boston: Houghton-Mifflin.
Carter, L. J. 1976. "National Environmental Policy Ac: Critics say Promise Unfulfilled." *Science* 193:130-132.
Changnon, S. A. 1976. "Inadvertent Weather Modification." *Water Resources Bulletin* 12(4):695-718.

Citizens Advisory Committee on Environmental Quality. 1974. *Report to President and to the Council on Environmental Quality.* Washington, D.C.: CACEQ.
Clawson, M. 1963. *Land and Water for Recreation.* Chicago: Rand McNally.
Copi, I. M. 1969, *Introduction to Logic,* 3d ed. New York: Macmillan.
Corwin, R., P. H. Heffernan, R. A. Johnston, et al. 1975. *Environmental Impact Assessment.* San Francisco: Freeman, Cooper.
Costle, D. M. "A Reveille, not a Requiem." Presented to the Energy and Environmental Policy Center, Harvard University, Cambridge, Mass., 1980.
Council on Environmental Quality. 1972. *Third Annual Report.* Washington, D.C.: Government Printing Office.
Council on Environmental Quality. 1973. "Preparation of Environmental Impact Statements: Guidelines." Federal Register 38:20550.
Council on Environmental Quality. 1975. *Environmental Quality. The Sixth Annual Report of the Council on Environmental Quality.* Washington, D.C.: U.S. Government Printing Office.
Council on Environmental Quality. 1976a. *Environmental Quality. The Seventh Annual Report of the Council on Environmental Quality.* Washington, D.C.: Government Printing Office.
Council on Environmental Quality. 1976b. *Environmental Impact Statements: An Analysis of Six Years' Experience by Seventy Federal Agencies.* Washington, D.C.: Government Printing Office.
Council on Environmental Quality. 1978. *Environmental Quality. The Ninth Annual Report.* Washington, D.C.: Government Printing Office.
Council on Environmental Quality. 1979. *Environmental Quality. The Tenth Annual Report.* Washington, D.C.: Government Printing Office.
Coyle, D.C. 1957. *Conservation.* New Brunswick, N.J.: Rutgers University Press.
Curlin, J. W. and H. S. Hughes. 1973. *National Environmental Policy Act of 1969: An Analysis of Proposed Legislative Modifications.* Washington, D.C.: Environmental Policy Division, Congressional Research Service, Library of Congress, Government Printing Office.
Dansereau, P. 1970. "Ecology and the Escalation of Human Impact." *International Social Science Journal* 22(4):628-647.
Dansereau, P. 1971. "Dimensions of Environmental Quality." *Sarracenia* 14:1-109.
Dansereau, P. 1973. *Inscape and Landscape.* Toronto: CBC Learning Systems.
Dee, N., et al. 1972. *Environmental Evaluation System for Water Resource Planning.* Columbus, O.: Battelle, Bureau of Reclamation, Columbus Laboratories.
Delong, R. L., W. G. Gilmartin, and J. G. Simpson, 1973. "Premature Births in California Sea Lions: Association with High Organochlorine Pollutant Levels." *Science* 181:1168-1170.
Environmental Protection Agency. 1973. *Environmental Impact Statement Guidelines.* Seattle, Wash.: EPA.

Environmental Protection Agency. 1975. *Manual: Review of Federal Actions Impacting the Environment.* Washington, D.C.: EPA.
Environmental Protection Agency. 1976. *Clean Air: The Breath of Life.* Washington, D.C.: EPA.
Esposito, J. C. 1970. *Vanishing Air.* New York: Grossman.
Graham, Frank, Jr. 1970. *Since Silent Spring.* Boston: Houghton Mifflin.
Green, M. J., J. M. Fallows and D. R. Zwick. 1972. *Who Runs Congress?* New York: Bantam.
Haik, R. A. 1974. "The Law and Environment Trade-offs." *American Forests* 80(8):18-20.
Harrington, W. 1981. The Endangered Species Act and the Search for Balance. *Natural Resources Journal* 21(1):71-92.
Hayakawa, S. I. 1974. How we Know What we Know. in *Speech Communication.* Edited by R. L. Applebaum, O. O. Jensen, and R. Carroll. New York: Macmillan.
Heer, J. E. and D. J. Hagerty. 1977. *Environmental Assessments and Statements.* New York: Van Nostrand Reinhold.
Henwood, K. and C. Coop. 1973. "Impact Analysis and the Planning Process." In *Environmental Quality and Water Development.* Edited by C. R. Goldman, J. McEvoy, P. J. Richerson. San Francisco: Freeman and Co.
Hill, J., H. P. Kollig, D. F. Paris, et al. 1976. *Dynamic Behavior of Vinyl Chloride in Aquatic Ecosystems.* EPA-600/3-76-001. Athens, Ga.: EPA.
Holmes, B. H. 1972. *A History of Water Resource Programs, 1800-1960.* Miscellaneous Publication No. 1233. U.S. Dept. of Agriculture Economic Research Service.
Holmes, B. H. 1979. *History of Water Resources Programs and Policies. 1961-1970.* Miscellaneous Publication No. 1379. Washington, D.C.: U.S. Dept. of Agriculture.
Hopkins, L. D. et al. 1973. *EIS: A Handbook for Writers and Reviewers.* NTIS 226-276. Urbana, Ill.: University of Illinois.
Hudson, D. R. 1974. "Environmental Management and Public Policy: An Analysis of the Environmental Impact Statement Process, with Emphasis on Procedures, etc." Doctoral dissertation, Georgia State University, Atlanta, Ga.
Jain, R. K. et al. 1974. *Handbook for Environmental Impact Analysis.* NTIS AD/A-006-241. Champaign, Ill: Construction Engineering Research Lab.
Krauskopf, T. M. and D. C. Bunde. 1972. "Evaluation of Environmental Impact Through a Computer Modelling Process." In *Environmental Impact Analysis: Philosophy and Methods.* Edited by R. B. Ditton and T. I. Goodale. Madison, Wisc.: University of Wisconsin Sea Grant Program.
League of Women Voters, 1972. *Anatomy of a Hearing.* Publ. No. 108. Washington, D.C.: LWV.
Leonardo Scholars. 1975. *Resources and Decisions.* North Scituate, Mass.: Duxbury Press.
Leopold, L. B. et al. 1971. A Procedure for Evaluating Environmental Impact.

Geological Survey Circular 645. Washington, D.C.: U.S. Dept. of the Interior.

Luken, R. and L. Langlois. 1973. "Innovations in Water Resource Planning: Creating and Communicating Discernable Alternatives." In *Environmental Quality and Water Development*. Edited by C. R. Goldman, J. McEvoy, and P. J. Richerson. San Francisco: Freeman and Co.

Mades, D. M. and G. Tauxe, 1980. Models and Methods in Multi-objective Water Resources Planning. Report No. 150. Water Resources Center, University of Illinois, Urbana, Ill.

Major, D. C. 1977. *Multi-objective Water Resource Planning*. Water Resources Monograph 4. Washington, D.C.: American Geophysical Union.

Manes, S. L. 1980. Alice in the Wonderland of SEQR. *New York State Bar Journal* 52(2):115.

Marsh, G. P. 1874. *The Earth as Modified by Human Action*. New York: Scribners.

Marx, L. 1970. "American Institutions and Ecological Ideals." *Science* 170:945-52.

McHarg, I. 1969. *Design With Nature*. Garden City, N.Y.: Natural History Press.

Moncrief, L. W. 1972. The Cultural Basis for Our Environmental Crisis. In *Economics of the Environment*. Edited by R. Dorfman and N. S. Dorfman. New York: Norton.

Munn, R. E. 1975. *Environmental Impact Assessment: Principles and Procedures*. SCOPE Report 5. Toronto.

Orloff, N. 1978. *The Environmental Impact Statement Process: A Guide to Citizen Action*. Washington, D.C.: Information Resources Press.

Peterson, R. W. 1976. "Editorial: The Impact Statement—Part 2." *Science* 193(4249).

Rosen, S. J. 1976. Manual for Environmental Impact Evaluation. Englewood Cliffs, N.J.: Prentice-Hall.

Ross, C. R. 1972. "Decision-making at Local, State, Federal, and International Levels." in *Environmental Quality and Water Development*. Edited by C. R. Goldman, J. McEvoy, and P. J. Richerson. San Francisco: Freeman and Company.

Russwurm, L. H. and E. Sommerville. 1974. Man's Natural Environment: A Systems Approach. North Scituate, Mass.; Duxbury Press.

Schad, T. M. 1968. "Congressional Handling of Water Resources." Congressional Record—House, Feb. 21, 1968. Washington, D.C.: Government Printing Office.

Schaeman, P. S. and T. Muller. 1974. *Measuring Impacts of Land Development: An Initial Approach*. Washington, D.C.: The Urban Institute.

Schindler, D. W. 1976. "The Impact Statement Boondoggle. *Science* 192:4239.

Senate Committee on Interior and Insular Affairs. 1968. Congressional Paper on a National Policy for the Environment. Serial T. Washington, D.C.: Government Printing Office.

Sewell, W. R. D. and T. O'Riordan. 1976. "The Culture of Participation in Environmental Decision-making." *Natural Resources Journal* 16(1):1-21.

Smith, J. N. 1974. "The Coming Age of Environmentalism in American Society." *Environmental Quality and Social Justice in Urban America*. Washington, D.C.: The Conservation Foundation.
Sorenson, J. C. 1971. "A Framework for Identification and Control of Resource Degradation and Conflict in the Multiple Use of the Coastal Zone." Masters' thesis, University of California, Berkeley.
Sorenson, J. C. 1972. "Some Procedures and Programs of Environmental Impact." In *Environmental Impact Analysis: Philosophy and Methods*. Edited by R. B. Ditton and T. I. Goodale. Madison, Wisc.: University of Wisconsin Sea Grant Program.
Spilhaus, A. 1972. "Ecolibrium." *Science* 175:711-715.
Stapp, W. 1975. Ecology, Man and His Environment—Public Pressure. University of Mich. Television, PBS Channel 24, Syracuse, New York, July 23.
Stellern, M. J. et al. 1979. *Environmental Quality Possibilities (E.Q.P.): A Procedure for Evaluating Economic/Environmental Tradeoffs*. Washington, D.C.: National Resources Economics Division, U.S. Dept. of Agriculture.
Stover, L. V. 1972. *Environmental Impact Assessment: A Procedure*. Miami, Fla.: Sanders and Thomas.
Subcommittee on Science, Research and Development, House Committee on Science and Astronautics. 1968. *Managing the Environment*. Congressional Record—House. Serial S. Washington, D.C.: Government Printing Office.
Sullivan, T. F. P. 1975. "National Environmental Policy Act," In *Environmental Law Handbook*. Government Institutes, Inc.
Train, R. E. 1974. *Ecology and Economy: Two Household Words*. Washington, D.C.: EPA.
Truman, D. B. 1959. *The Governmental Process*. New York: Knopf.
Trzyna, T. C. 1974. *Environmental Impact Requirements in the States: N.E.P.A.'s Offspring*. EPA No. 68-01-1818. Washington, D.C.: Government Printing Office.
Udall, S. L. 1963. *The Quiet Crisis*. New York: Holt, Rinehart and Winston.
Urban Land Institute. *Environment and the Land Developer*. Washington, D.C.: ULI, 1971.
U.S. Forest Service. 1952. *Highlights in the History of Forest Conservation*. Agriculture Information Bulletin No. 83. Washington, D.C.: Government Printing Office.
Wandesforde-Smith, G. 1971. "The Bureaucratic Response to Environmental Politics." *Natural Resources Journal* 11(3):479-488.
Water Resources Council. 1962. Policies, Standards, and Procedures in the Formulation, Evaluation, and Review of Plans of Use and Development of Water and Related Land Resources. Senate Document No. 97, 87th Cong., 2d Sess.
Watson, J. D. 1968. *The Double Helix*. New York: Atheneum Press.
Wengert, N. 1976. "Citizen Participation: Practice in Search of a Theory." *Natural Resources Journal* 16(1): 22-40.

White, G. F. 1972a. "Environmental Impact Statements." *Professional Geographer* 24(4):306.

White, G. F. 1972b. "Public Opinion in Planning Water Development." In *Environmental Quality and Water Development*. Edited by C. R. Goldman, J. McEvoy, and P. J. Richerson. San Francisco: Freeman and Co.

Wichelman, A. F. 1976. "Administrative Agency Implementation of the National Environmental Policy Act of 1969: A Conceptual Framework for Explaining Differential Response." *Natural Resources Journal* 16(2):263-300.

Wilmot, E. L. and R. E. Luna. 1980. *Worst-Case Scenario Syndrome*. Albuquerque, N. M.: Sandia Labs.

Wurster, C. F. and D. B. Wingate. 1968. "DDT Residues and Declining Reproduction in the Bermuda Petrel." *Science* 149:979-981.

Yorke, T. H. 1978. Impact Assessment of Water Resource Development Activities: a Dual Matrix Approach. FWS/OBS 78/82. Washington, D.C.: Fish and Wildlife Service, U.S. Dept. of the Interior.

Appendixes

A. The National Environmental Policy Act of 1969, as Amended*
B. The Environmental Quality Improvement Act of 1970*
C. CEQ Regulations for Implementing the Procedural Provisions of the National Environmental Policy Act*
D. The Clean Air Act §309*
E. Executive Order 11514 (as amended by Executive Order 11911) and Executive Order 12114*
F. Relationships Between NEPA Requirements for EIS Contents and the Requirements of Procedures for Evaluation of Environmental Quality Objective

---

*Reproduced from: Council of Environmental Quality. *Regulations for Implementing the Procedural Provisions of the National Environmental Policy Act*. Reprint No. 43 FR 55978-56007. Washington, D.C.: Government Printing Office, 1978.

APPENDIX A: The National Environmental Policy Act of 1969, as Amended

An Act to establish a national policy for the environment, to provide for the establishment of a Council on Environmental Quality, and for other purposes.

Be it enacted by the Senate and House of Representatives of the United States of America in Congress assembled, That this Act may be cited as the "National Environmental Policy Act of 1969."

PURPOSE

SEC. 2. The purposes of this Act are: To declare a national policy which will encourage productive and enjoyable harmony between man and his environment; to promote efforts which will prevent or eliminate damage to the environment and biosphere and stimulate the health and welfare of man; to enrich the understanding of the ecological systems and natural resources important to the Nation; and to establish a Council on Environmental Quality.

TITLE I

DECLARATION OF NATIONAL ENVIRONMENTAL POLICY

SEC. 101. (a) The Congress, recognizing the profound impact of man's activity on the interrelations of all components of the natural environment, particularly the profound influences of population growth, high-density urbanization, industrial expansion, resource exploitation, and new and expanding technological advances and recognizing further the critical importance of restoring and maintaining environmental quality to the overall welfare and development of man, declares that it is the continuing policy of the Federal Government, in cooperation with State and local governments, and other concerned public and private organizations, to use all practicable means and measures, including financial and technical assistance, in a manner calculated to foster and promote the general welfare, to create and maintain conditions under which man and nature can exist in productive harmony, and fulfill the social, economic, and other requirements of present and future generations of Americans.

(b) In order to carry out the policy set forth in this Act, it is the continuing responsibility of the Federal Government to use all practicable means, consistent with other essential considerations of national policy, to improve and coordinate Federal plans, functions, programs, and resources to the end that the Nation may—

(1) fulfill the responsibilities of each generation as trustee of the environment for succeeding generations;

(2) assure for all Americans safe, healthful, productive, and esthetically and culturally pleasing surroundings;

Pub. L. 91–190, 42 U.S.C. 4321–4347, January 1, 1970, as amended by Pub. L. 94–52, July 3, 1975, and Pub. L. 94–83, August 9, 1975.

(3) attain the widest range of beneficial uses of the environment without degradation, risk to health or safety, or other undesirable and unintended consequences;

(4) preserve important historic, cultural, and natural aspects of our national heritage, and maintain, wherever possible, an environment which supports diversity, and variety of individual choice;

(5) achieve a balance between population and resource use which will permit high standards of living and a wide sharing of life's amenities; and

(6) enhance the quality of renewable resources and approach the maximum attainable recycling of depletable resources.

(c) The Congress recognizes that each person should enjoy a healthful environment and that each person has a responsibility to contribute to the preservation and enhancement of the environment.

Sec. 102. The Congress authorizes and directs that, to the fullest extent possible: (1) the policies, regulations, and public laws of the United States shall be interpreted and administered in accordance with the policies set forth in this Act, and (2) all agencies of the Federal Government shall—

(A) Utilize a systematic, interdisciplinary approach which will insure the integrated use of the natural and social sciences and the environmental design arts in planning and in decisionmaking which may have an impact on man's environment;

(B) Identify and develop methods and procedures, in consultation with the Council on Environmental Quality established by title II of this Act, which will insure that presently unquantified environmental amenities and values may be given appropriate consideration in decisionmaking along with economic and technical considerations;

(C) Include in every recommendation or report on proposals for legislation and other major Federal actions significantly affecting the quality of the human environment, a detailed statement by the responsible official on—

(i) The environmental impact of the proposed action,

(ii) Any adverse environmental effects which cannot be avoided should the proposal be implemented,

(iii) Alternatives to the proposed action,

(iv) The relationship between local short-term uses of man's environment and the maintenance and enhancement of long-term productivity, and

(v) Any irreversible and irretrievable commitments of resources which would be involved in the proposed action should it be implemented.

Prior to making any detailed statement, the responsible Federal official shall consult with and obtain the comments of any Federal agency which has jurisdiction by law or special expertise with respect to any environmental impact involved. Copies of such statement and the comments and views of the appropriate Federal, State, and local agencies, which are authorized to develop and enforce environmental standards, shall be made available to the President, the Council on Environmental Quality and to the public as provided by section 552 of title 5, United States Code, and shall accompany the proposal through the existing agency review processes;

(d) Any detailed statement required under subparagraph (c) after January 1, 1970, for any major Federal action funded under a program of grants to States shall not be deemed to be legally insufficient solely by reason of having been prepared by a State agency or official, if:

 (i) the State agency or official has statewide jurisdiction and has the responsibility for such action,

 (ii) the responsible Federal official furnishes guidance and participates in such preparation,

 (iii) the responsible Federal official independently evaluates such statement prior to its approval and adoption, and

 (iv) after January 1, 1976, the responsible Federal official provides early notification to, and solicits the views of, any other State or any Federal land management entity of any action or any alternative thereto which may have significant impacts upon such State or affected Federal land management entity and, if there is any disagreement on such impacts, prepares a written assessment of such impacts and views for incorporation into such detailed statement.

The procedures in this subparagraph shall not relieve the Federal official of his responsibilities for the scope, objectivity, and content of the entire statement or of any other responsibility under this Act; and further, this subparagraph does not affect the legal sufficiency of statements prepared by State agencies with less than statewide jurisdiction.

(e) Study, develop, and describe appropriate alternatives to recommended courses of action in any proposal which involves unresolved conflicts concerning alternative uses of available resources;

(f) Recognize the worldwide and long-range character of environmental problems and, where consistent with the foreign policy of the United States, lend appropriate support to initiatives, resolutions, and programs designed to maximize international cooperation in anticipating and preventing a decline in the quality of mankind's world environment;

(g) Make available to States, counties, municipalities, institutions, and individuals, advice and information useful in restoring, maintaining, and enhancing the quality of the environment;

(h) Initiate and utilize ecological information in the planning and development of resource-oriented projects; and

(i) Assist the Council on Environmental Quality established by title II of this Act.

Sec. 103. All agencies of the Federal Government shall review their present statutory authority, administrative regulations, and current policies and procedures for the purpose of determining whether there are any deficiencies or inconsistencies therein which prohibit full compliance with the purposes and provisions of this Act and shall propose to the President not later than July 1, 1971, such measures as may be necessary to bring their authority and policies into conformity with the intent, purposes, and procedures set forth in this Act.

Sec. 104. Nothing in section 102 or 103 shall in any way affect the specific statutory obligations of any Federal agency (1) to comply with criteria or standards of environmental quality, (2) to coordinate or consult with any other Federal or State agency, or (3) to act, or refrain from acting contingent upon the recommendations or certification of any other Federal or State agency.

Sec. 105. The policies and goals set forth in this Act are supplementary to those set forth in existing authorizations of Federal agencies.

TITLE II

COUNCIL ON ENVIRONMENTAL QUALITY

SEC. 201. The President shall transmit to the Congress annually beginning July 1, 1970, an Environmental Quality Report (hereinafter referred to as the "report") which shall set forth (1) the status and condition of the major natural, manmade, or altered environmental classes of the Nation, including, but not limited to, the air, the aquatic, including marine, estuarine, and fresh water, and the terrestrial environment, including, but not limited to, the forest, dryland, wetland, range, urban, suburban and rural environment; (2) current and foreseeable trends in the quality, management and utilization of such environments and the effects of those trends on the social, economic, and other requirements of the Nation; (3) the adequacy of available natural resources for fulfilling human and economic requirements of the Nation in the light of expected population pressures; (4) a review of the programs and activities (including regulatory activities) of the Federal Government, the State and local governments, and nongovernmental entities or individuals with particular reference to their effect on the environment and on the conservation, development and utilization of natural resources; and (5) a program for remedying the deficiencies of existing programs and activities, together with recommendations for legislation.

SEC. 202. There is created in the Executive Office of the President a Council on Environmental Quality (hereinafter referred to as the "Council"). The Council shall be composed of three members who shall be appointed by the President to serve at his pleasure, by and with the advice and consent of the Senate. The President shall designate one of the members of the Council to serve as Chairman. Each member shall be a person who, as a result of his training, experience, and attainments, is exceptionally well qualified to analyze and interpret environmental trends and information of all kinds; to appraise programs and activities of the Federal Government in the light of the policy set forth in title I of this Act; to be conscious of and responsive to the scientific, economic, social, esthetic, and cultural needs and interests of the Nation; and to formulate and recommend national policies to promote the improvement of the quality of the environment.

SEC. 203. The Council may employ such officers and employees as may be necessary to carry out its functions under this Act. In addition, the Council may employ and fix the compensation of such experts and consultants as may be necessary for the carrying out of its functions under this Act, in accordance with section 3109 of title 5, United States Code (but without regard to the last sentence thereof).

SEC. 204. It shall be the duty and function of the Council—

(1) to assist and advise the President in the preparation of the Environmental Quality Report required by section 201 of this title;

(2) to gather timely and authoritative information concerning the conditions and trends in the quality of the environment both current and prospective, to analyze and interpret such information for the purpose of determining whether such conditions and trends are interfering, or are likely to interfere, with the achievement of the policy set forth in title I of this Act, and to compile and submit to the President studies relating to such conditions and trends;

(3) to review and appraise the various programs and activities of the Federal Government in the light of the policy set forth in title I of this

Act for the purpose of determining the extent to which such programs and activities are contributing to the achievement of such policy, and to make recommendations to the President with respect thereto;

(4) to develop and recommend to the President national policies to foster and promote the improvement of environmental quality to meet the conservation, social, economic, health, and other requirements and goals of the Nation;

(5) to conduct investigations, studies, surveys, research, and analyses relating to ecological systems and environmental quality;

(6) to document and define changes in the natural environment, including the plant and animal systems, and to accumulate necessary data and other information for a continuing analysis of these changes or trends and an interpretation of their underlying causes;

(7) to report at least once each year to the President on the state and condition of the environment; and

(8) to make and furnish such studies, reports thereon, and recommendations with respect to matters of policy and legislation as the President may request.

SEC. 205. In exercising its powers, functions, and duties under this Act, the Council shall—

(1) Consult with the Citizens' Advisory Committee on Environmental Quality established by Executive Order No. 11472, dated May 29, 1969, and with such representatives of science, industry, agriculture, labor, conservation organizations, State and local governments and other groups, as it deems advisable; and

(2) Utilize, to the fullest extent possible, the services, facilities and information (including statistical information) of public and private agencies and organizations, and individuals, in order that duplication of effort and expense may be avoided, thus assuring that the Council's activities will not unnecessarily overlap or conflict with similar activities authorized by law and performed by established agencies.

SEC. 206. Members of the Council shall serve full time and the Chairman of the Council shall be compensated at the rate provided for Level II of the Executive Schedule Pay Rates (5 U.S.C. 5313). The other members of the Council shall be compensated at the rate provided for Level IV of the Executive Schedule Pay Rates (5 U.S.C. 5315).

SEC. 207. The Council may accept reimbursements from any private nonprofit organization or from any department, agency, or instrumentality of the Federal Government, any State, or local government, for the reasonable travel expenses incurred by an officer or employee of the Council in connection with his attendance at any conference, seminar, or similar meeting conducted for the benefit of the Council.

SEC. 208. The Council may make expenditures in support of its international activities, including expenditures for: (1) international travel; (2) activities in implementation of international agreements; and (3) the support of international exchange programs in the United States and in foreign countries.

SEC. 209. There are authorized to be appropriated to carry out the provisions of this chapter not to exceed $300,000 for fiscal year 1970, $700,000 for fiscal year 1971, and $1,000,000 for each fiscal year thereafter.

APPENDIX B: The Environmental Quality Improvement Act of 1970

TITLE II—ENVIRONMENTAL QUALITY
(OF THE WATER QUALITY IMPROVEMENT ACT OF 1974)

SHORT TITLE

SEC. 201. This title may be cited as the "Environmental Quality Improvement Act of 1970."

FINDINGS, DECLARATIONS, AND PURPOSES

SEC. 202. (a) The Congress finds—
   (1) That man has caused changes in the environment;
   (2) That many of these changes may affect the relationship between man and his environment; and
   (3) That population increases and urban concentration contribute directly to pollution and the degradation of our environment.

(b)(1) The Congress declares that there is a national policy for the environment which provides for the enhancement of environmental quality. This policy is evidenced by statutes heretofore enacted relating to the prevention, abatement, and control of environmental pollution, water and land resources, transportation, and economic and regional development.

(2) The primary responsibility for implementing this policy rests with State and local governments.

(3) The Federal Government encourages and supports implementation of this policy through appropriate regional organizations established under existing law.

(c) The purposes of this title are—
   (1) To assure that each Federal department and agency conducting or supporting public works activities which affect the environment shall implement the policies established under existing law; and
   (2) To authorize an Office of Environmental Quality, which, notwithstanding any other provision of law, shall provide the professional and administrative staff for the Council on Environmental Quality established by Public Law 91-190.

OFFICE OF ENVIRONMENTAL QUALITY

SEC. 203. (a) There is established in the Executive Office of the President an office to be known as the Office of Environmental Quality (hereafter in this title referred to as the "Office"). The Chairman of the Council on Environmental Quality established by Public Law 91-190 shall be the Director of the Office. There shall be in the Office a Deputy Director who shall be appointed by the President, by and with the advice and consent of the Senate.

(b) The compensation of the Deputy Director shall be fixed by the President at a rate not in excess of the annual rate of compensation payable to the Deputy Director of the Bureau of the Budget.

Pub. L. 91-224, 42 U.S.C. 4371-4374, April 3, 1970.

(c) The Director is authorized to employ such officers and employees (including experts and consultants) as may be necessary to enable the Office to carry out its functions under this title and Public Law 91-190, except that he may employ no more than 10 specialists and other experts without regard to the provisions of title 5, United States Code, governing appointments in the competitive service, and pay such specialists and experts without regard to the provisions of chapter 51 and subchapter 111 of chapter 53 of such title relating to classification and General Schedule pay rates, but no such specialist or expert shall be paid at a rate in excess of the maximum rate for GS-18 of the General Schedule under section 5330 of title 5.

(d) In carrying out his functions the Director shall assist and advise the President on policies and programs of the Federal Government affecting environmental quality by—

(1) Providing the professional and administrative staff and support for the Council on Environmental Quality established by Public Law 91-190;

(2) Assisting the Federal agencies and departments in appraising the effectiveness of existing and proposed facilities, programs, policies, and activities of the Federal Government, and those specific major projects designated by the President which do not require individual project authorization by Congress, which affect environmental quality;

(3) Reviewing the adequacy of existing systems for monitoring and predicting environmental changes in order to achieve effective coverage and efficient use of research facilities and other resources;

(4) Promoting the advancement of scientific knowledge of the effects of actions and technology on the environment and encourage the development of the means to prevent or reduce adverse effects that endanger the health and well-being of man;

(5) Assisting in coordinating among the Federal departments and agencies those programs and activities which affect, protect, and improve environmental quality;

(6) Assisting the Federal departments and agencies in the development and interrelationship of environmental quality criteria and standards established through the Federal Government;

(7) Collecting, collating, analyzing, and interpreting data and information on environmental quality, ecological research, and evaluation.

(e) The Director is authorized to contract with public or private agencies, institutions, and organizations and with individuals without regard to sections 3618 and 3709 of the Revised Statutes (31 U.S.C. 529; 41 U.S.C. 5) in carrying out his functions.

REPORT

SEC. 204. Each Environmental Quality Report required by Public Law 91-190 shall, upon transmittal to Congress, be referred to each standing committee having jurisdiction over any part of the subject matter of the Report.

AUTHORIZATION

SEC. 205. There are hereby authorized to be appropriated not to exceed $500,000 for the fiscal year ending June 30, 1970, not to exceed $750,000 for the fiscal year ending June 30, 1971, not to exceed $1,250,000 for the fiscal year ending June 30, 1972, and not to exceed $1,500,000 for the fiscal year ending June 30, 1973. These authorizations are in addition to those contained in Public Law 91-190.

Approved April 3, 1970.

APPENDIX C: Regulations for Implementing the Procedural Provisions of the National Environmental Policy Act

PART 1500—PURPOSE, POLICY, AND MANDATE

1500.1 Purpose.
1500.2 Policy.
1500.3 Mandate.
1500.4 Reducing paperwork.
1500.5 Reducing delay.
1500.6 Agency authority.

AUTHORITY: NEPA, the Environmental Quality Improvement Act of 1970, as amended (42 U.S.C. 4371 et seq.), section 309 of the Clean Air Act, as amended (42 U.S.C. 7609) and Executive Order 11514, Protection and Enhancement of Environmental Quality (March 5, 1970 as amended by Executive Order 11991, May 24, 1977).

§ 1500.1 Purpose.

(a) The National Environmental Policy Act (NEPA) is our basic national charter for protection of the environment. It establishes policy, sets goals (section 101), and provides means (section 102) for carrying out the policy. Section 102(2) contains "action-forcing" provisions to make sure that federal agencies act according to the letter and spirit of the Act. The regulations that follow implement Section 102(2). Their purpose is to tell federal agencies what they must do to comply with the procedures and achieve the goals of the Act. The President, the federal agencies, and the courts share responsibility for enforcing the Act so as to achieve the substantive requirements of section 101.

(b) NEPA procedures must insure that environmental information is available to public officials and citizens before decisions are made and before actions are taken. The information must be of high quality. Accurate scientific analysis, expert agency comments, and public scrutiny are essential to implementing NEPA. Most important, NEPA documents must concentrate on the issues that are truly significant to the action in question, rather than amassing needless detail.

(c) Ultimately, of course, it is not better documents but better decisions that count. NEPA's purpose is not to generate paperwork—even excellent paperwork—but to foster excellent action. The NEPA process is intended to help public officials make decisions that are based on understanding of environmental consequences, and take actions that protect, restore, and enhance the environment. These regulations provide the direction to achieve this purpose.

§ 1500.2 Policy.

Federal agencies shall to the fullest extent possible:

(a) Interpret and administer the policies, regulations, and public laws of the United States in accordance with the policies set forth in the Act and in these regulations.

(b) Implement procedures to make the NEPA process more useful to decisionmakers and the public; to reduce paperwork and the accumulation of extraneous background data; and to emphasize real environmental issues and alternatives. Environmental impact statements shall be concise, clear, and to the point, and shall be supported by evidence that agencies have made the necessary environmental analyses.

(c) Integrate the requirements of NEPA with other planning and environmental review procedures required by law or by agency practice so that all such procedures run concurrently rather than consecutively.

(d) Encourage and facilitate public involvement in decisions which affect the quality of the human environment.

(e) Use the NEPA process to identify and assess the reasonable alternatives to proposed actions that will avoid or minimize adverse effects of these actions upon the quality of the human environment.

(f) Use all practicable means, consistent with the requirements of the Act and other essential considerations of national policy, to restore and enhance the quality of the human environment and avoid or minimize any possible adverse effects of their actions upon the quality of the human environment.

§ 1500.3 Mandate.

Parts 1500-1508 of this Title provide regulations applicable to and binding on all Federal agencies for implementing the procedural provisions of the National Environmental Policy Act of 1969, as amended (Pub. L. 91-190, 42 U.S.C. 4321 et seq.) (NEPA or the Act) except where compliance would be inconsistent with other statutory requirements. These regulations are issued pursuant to NEPA, the Environmental Quality Improvement Act of 1970, as amended (42 U.S.C. 4371 et seq.) Section 309 of the Clean Air Act, as amended (42 U.S.C. 7609) and Executive Order 11514, Protection and Enhancement of Environmental Quality (March 5, 1970, as amended by Executive Order 11991, May 24, 1977). These regulations, unlike the predecessor guidelines, are not confined to Sec. 102(2)(C) (environmental impact statements). The regulations apply to the whole of section 102(2). The provisions of the Act and of these regulations must be read together as a whole in order to comply with the spirit and letter of the law. It is the Council's intention that judicial review of agency compliance with these regulations not occur before an agency has filed the final environmental impact statement, or has made a final finding of no significant impact (when such a finding will result in action affecting the environment), or takes action that will result in irreparable injury. Furthermore, it is the Council's intention that any trivial violation of these regulations not give rise to any independent cause of action.

§ 1500.4 Reducing paperwork.

Agencies shall reduce excessive paperwork by:

(a) Reducing the length of environmental impact statements (§ 1502.2(c)), by means such as setting appropriate page limits (§§ 1501.7(b)(1) and 1502.7).

(b) Preparing analytic rather than encyclopedic environmental impact statements (§ 1502.2(a)).

(c) Discussing only briefly issues other than significant ones (§ 1502.2(b)).

(d) Writing environmental impact statements in plain language (§ 1502.8).

(e) Following a clear format for environmental impact statements (§ 1502.10).

(f) Emphasizing the portions of the environmental impact statement that are useful to decisionmakers and the public (§§ 1502.14 and 1502.15) and reducing emphasis on background material (§ 1502.16).

(g) Using the scoping process, not only to identify significant environmental issues deserving of study, but also to deemphasize insignificant issues, narrowing the scope of the environmental impact statement process accordingly (§ 1501.7).

(h) Summarizing the environmental impact statement (§ 1502.12) and circulating the summary instead of the entire environmental impact statement if the latter is unusually long (§ 1502.19).

(i) Using program, policy, or plan environmental impact statements and tiering from statements of broad scope to those of narrower scope, to eliminate repetitive discussions of the same issues (§§ 1502.4 and 1502.20).

(j) Incorporating by reference (§ 1502.21).

(k) Integrating NEPA requirements with other environmental review and consultation requirements (§ 1502.25).

(l) Requiring comments to be as specific as possible (§ 1503.3).

(m) Attaching and circulating only changes to the draft environmental impact statement, rather than rewriting and circulating the entire statement when changes are minor (§ 1503.4(c)).

(n) Eliminating duplication with State and local procedures, by providing for joint preparation (§ 1506.2), and with other Federal procedures, by providing that an agency may adopt appropriate environmental documents prepared by another agency (§ 1506.3).

(o) Combining environmental documents with other documents (§ 1506.4).

(p) Using categorical exclusions to define categories of actions which do not individually or cumulatively have a significant effect on the human environment and which are therefore exempt from requirements to prepare an environmental impact statement (§ 1508.4).

(q) Using a finding of no significant impact when an action not otherwise excluded will not have a significant effect on the human environment and is therefore exempt from requirements to prepare an environmental impact statement (§ 1508.13).

§ 1500.5 Reducing delay.

Agencies shall reduce delay by:
(a) Integrating the NEPA process into early planning (§ 1501.2).
(b) Emphasizing interagency cooperation before the environmental impact statement is prepared, rather than submission of adversary comments on a completed document (§ 1501.6).
(c) Insuring the swift and fair resolution of lead agency disputes (§ 1501.5).
(d) Using the scoping process for an early identification of what are and what are not the real issues (§ 1501.7).
(e) Establishing appropriate time limits for the environmental impact statement process (§§ 1501.7(b)(2) and 1501.8).
(f) Preparing environmental impact statements early in the process (§ 1502.5).
(g) Integrating NEPA requirements with other environmental review and consultation requirements (§ 1502.25).
(h) Eliminating duplication with State and local procedures by providing for joint preparation (§ 1506.2) and with other Federal procedures by providing that an agency may adopt appropriate environmental documents prepared by another agency (§ 1506.3).
(i) Combining environmental documents with other documents (§ 1506.4).
(j) Using accelerated procedures for proposals for legislation (§ 1506.8).
(k) Using categorical exclusions to define categories of actions which do not individually or cumulatively have a significant effect on the human environment (§ 1508.4) and which are therefore exempt from requirements to prepare an environmental impact statement.
(1) Using a finding of no significant impact when an action not otherwise excluded will not have a significant effect on the human environment (§ 1508.13) and is therefore exempt from requirements to prepare an environmental impact statement.

§ 1500.6 Agency authority.

Each agency shall interpret the provisions of the Act as a supplement to its existing authority and as a mandate to view traditional policies and missions in the light of the Act's national environmental objectives. Agencies shall review their policies, procedures, and regulations accordingly and revise them as necessary to insure full compliance with the purposes and provisions of the Act. The phrase "to the fullest extent possible" in section 102 means that each agency of the Federal Government shall comply with that section unless existing law applicable to the agency's operations expressly prohibits or makes compliance impossible.

PART 1501—NEPA AND AGENCY PLANNING

Sec.
1501.1 Purpose.
1501.2 Apply NEPA early in the process.

Sec.
1501.3 When to prepare an environmental assessment.
1501.4 Whether to prepare an environmental impact statement.
1501.5 Lead agencies.
1501.6 Cooperating agencies.
1501.7 Scoping.
1501.8 Time limits.

AUTHORITY: NEPA, the Environmental Quality Improvement Act of 1970, as amended (42 U.S.C. 4371 et seq.), Section 309 of the Clean Air Act, as amended (42 U.S.C. 7609), and Executive Order 11514, Protection and Enhancement of Environmental Quality (March 5, 1970, as amended by Executive Order 11991, May 24 1977).

§ 1501.1 Purpose.

The purposes of this part include:

(a) Integrating the NEPA process into early planning to insure appropriate consideration of NEPA's policies and to eliminate delay.

(b) Emphasizing cooperative consultation among agencies before the environmental impact statement is prepared rather than submission of adversary comments on a completed document.

(c) Providing for the swift and fair resolution of lead agency disputes.

(d) Identifying at an early stage the significant environmental issues deserving of study and deemphasizing insignificant issues, narrowing the scope of the environmental impact statement accordingly.

(e) Providing a mechanism for putting appropriate time limits on the environmental impact statement process.

§ 1501.2 Apply NEPA early in the process.

Agencies shall integrate the NEPA process with other planning at the earliest possible time to insure that planning and decisions reflect environmental values, to avoid delays later in the process, and to head off potential conflicts. Each agency shall:

(a) Comply with the mandate of section 102(2)(A) to "utilize a systematic, interdisciplinary approach which will insure the integrated use of the natural and social sciences and the environmental design arts in planning and in decisionmaking which may have an impact on man's environment," as specified by § 1507.2.

(b) Identify environmental effects and values in adequate detail so they can be compared to economic and technical analyses. Environmental documents and appropriate analyses shall be circulated and reviewed at the same time as other planning documents.

(c) Study, develop, and describe appropriate alternatives to recommended courses of action in any proposal which involves unresolved conflicts concerning alternative uses of available resources as provided by section 102(2)(E) of the Act.

(d) Provide for cases where actions are planned by private applicants or other non-Federal entities before Federal involvement so that:

(1) Policies or designated staff are available to advise potential applicants of studies or other information foreseeably required for later Federal action.

(2) The Federal agency consults early with appropriate State and local agencies and Indian tribes and with interested private persons and organizations when its own involvement is reasonably foreseeable.

(3) The Federal agency commences its NEPA process at the earliest possible time.

§ 1501.3 When to prepare an environmental assessment.

(a) Agencies shall prepare an environmental assessment (§ 1508.9) when necessary under the procedures adopted by individual agencies to supplement these regulations as described in § 1507.3. An assessment is not necessary if the agency has decided to prepare an environmental impact statement.

(b) Agencies may prepare an environmental assessment on any action at any time in order to assist agency planning and decisionmaking.

§ 1501.4 Whether to prepare an environmental impact statement.

In determining whether to prepare an environmental impact statement the Federal agency shall:

(a) Determine under its procedures supplementing these regulations (described in § 1507.3) whether the proposal is one which:
(1) Normally requires an environmental impact statement, or
(2) Normally does not require either an environmental impact statement or an environmental assessment (categorical exclusion).
(b) If the proposed action is not covered by paragraph (a) of this section, prepare an environmental assessment (§ 1508.9). The agency shall involve environmental agencies, applicants, and the public, to the extent practicable, in preparing assessments required by § 1508.9(a)(1).
(c) Based on the environmental assessment make its determination whether to prepare an environmental impact statement.
(d) Commence the scoping process (§ 1501.7), if the agency will prepare an environmental impact statement.
(e) Prepare a finding of no significant impact (§ 1508.13), if the agency determines on the basis of the environmental assessment not to prepare a statement.
(1) The agency shall make the finding of no significant impact available to the affected public as specified in § 1506.6.
(2) In certain limited circumstances, which the agency may cover in its procedures under § 1507.3, the agency shall make the finding of no significant impact available for public review (including State and areawide clearinghouses) for 30 days before the agency makes its final determination whether to prepare an environmental impact statement and before the action may begin. The circumstances are:
(i) The proposed action is, or is closely similar to, one which normally requires the preparation of an environmental impact statement under the procedures adopted by the agency pursuant to § 1507.3, or
(ii) The nature of the proposed action is one without precedent.

§ 1501.5 Lead agencies.

(a) A lead agency shall supervise the preparation of an environmental impact statement if more than one Federal agency either:
(1) Proposes or is involved in the same action; or
(2) Is involved in a group of actions directly related to each other because of their functional interdependence or geographical proximity.
(b) Federal, State, or local agencies, including at least one Federal agency, may act as joint lead agencies to prepare an environmental impact statement (§ 1506.2).
(c) If an action falls within the provisions of paragraph (a) of this section the potential lead agencies shall determine by letter or memorandum which agency shall be the lead agency and which shall be cooperating agencies. The agencies shall resolve the lead agency question so as not to cause delay. If there is disagreement among the agencies, the following factors (which are listed in order of descending importance) shall determine lead agency designation:
(1) Magnitude of agency's involvement.
(2) Project approval/disapproval authority.
(3) Expertise concerning the action's environmental effects.
(4) Duration of agency's involvement.
(5) Sequence of agency's involvement.
(d) Any Federal agency, or any State or local agency or private person substantially affected by the absence of lead agency designation, may make a written request to the potential lead agencies that a lead agency be designated.
(e) If Federal agencies are unable to agree on which agency will be the lead agency or if the procedure described in paragraph (c) of this section has not resulted within 45 days in a lead agency designation, any of the agencies or persons concerned may file a request with the Council asking it to determine which Federal agency shall be the lead agency.
A copy of the request shall be transmitted to each potential lead agency. The request shall consist of:
(1) A precise description of the nature and extent of the proposed action.

(2) A detailed statement of why each potential lead agency should or should not be the lead agency under the criteria specified in paragraph (c) of this section.

(f) A response may be filed by any potential lead agency concerned within 20 days after a request is filed with the Council. The Council shall determine as soon as possible but not later than 20 days after receiving the request and all responses to it which Federal agency shall be the lead agency and which other Federal agencies shall be cooperating agencies.

§ 1501.6 Cooperating agencies.

The purpose of this section is to emphasize agency cooperation early in the NEPA process. Upon request of the lead agency, any other Federal agency which has jurisdiction by law shall be a cooperating agency. In addition any other Federal agency which has special expertise with respect to any environmental issue, which should be addressed in the statement may be a cooperating agency upon request of the lead agency. An agency may request the lead agency to designate it a cooperating agency.

(a) The lead agency shall:

(1) Request the participation of each cooperating agency in the NEPA process at the earliest possible time.

(2) Use the environmental analysis and proposals of cooperating agencies with jurisdiction by law or special expertise, to the maximum extent possible consistent with its responsibility as lead agency.

(3) Meet with a cooperating agency at the latter's request.

(b) Each cooperating agency shall:

(1) Participate in the NEPA process at the earliest possible time.

(2) Participate in the scoping process (described below in § 1501.7).

(3) Assume on request of the lead agency responsibility for developing information and preparing environmental analyses including portions of the environmental impact statement concerning which the cooperating agency has special expertise.

(4) Make available staff support at the lead agency's request to enhance the latter's interdisciplinary capability.

(5) Normally use its own funds. The lead agency shall, to the extent available funds permit, fund those major activities or analyses it requests from cooperating agencies. Potential lead agencies shall include such funding requirements in their budget requests.

(c) A cooperating agency may in response to a lead agency's request for assistance in preparing the environmental impact statement (described in paragraph (b) (3), (4), or (5) of this section) reply that other program commitments preclude any involvement or the degree of involvement requested in the action that is the subject of the environmental impact statement. A copy of this reply shall be submitted to the Council.

§ 1501.7 Scoping.

There shall be an early and open process for determining the scope of issues to be addressed and for identifying the significant issues related to a proposed action. This process shall be termed scoping. As soon as practicable after its decision to prepare an environmental impact statement and before the scoping process the lead agency shall publish a notice of intent (§ 1508.22) in the FEDERAL REGISTER except as provided in § 1507.3(e).

(a) As part of the scoping process the lead agency shall:

(1) Invite the participation of affected Federal, State, and local agencies, any affected Indian tribe, the proponent of the action, and other interested persons (including those who might not be in accord with the action on environmental grounds), unless there is a limited exception under § 1507.3(c). An agency may give notice in accordance with § 1506.6.

(2) Determine the scope (§ 1508.25) and the significant issues to be analyzed in depth in the environmental impact statement.

(3) Identify and eliminate from detailed study the issues which are not significant or which have been covered by prior environmental review

(§ 1506.3), narrowing the discussion of these issues in the statement to a brief presentation of why they will not have a significant effect on the human environment or providing a reference to their coverage elsewhere.

(4) Allocate assignments for preparation of the environmental impact statement among the lead and cooperating agencies, with the lead agency retaining responsibility for the statement.

(5) Indicate any public environmental assessments and other environmental impact statements which are being or will be prepared that are related to but are not part of the scope of the impact statement under consideration.

(6) Identify other environmental review and consultation requirements so the lead and cooperating agencies may prepare other required analyses and studies concurrently with, and integrated with, the environmental impact statement as provided in § 1502.25.

(7) Indicate the relationship between the timing of the preparation of environmental analyses and the agency's tentative planning and decisionmaking schedule.

(b) As part of the scoping process the lead agency may:

(1) Set page limits on environmental documents (§ 1502.7).

(2) Set time limits (§ 1501.8).

(3) Adopt procedures under § 1507.3 to combine its environmental assessment process with its scoping process.

(4) Hold an early scoping meeting or meetings which may be integrated with any other early planning meeting the agency has. Such a scoping meeting will often be appropriate when the impacts of a particular action are confined to specific sites.

(c) An agency shall revise the determinations made under paragraphs (a) and (b) of this section if substantial changes are made later in the proposed action, or if significant new circumstances or information arise which bear on the proposal or its impacts.

§ 1501.8   Time limits.

Although the Council has decided that prescribed universal time limits for the entire NEPA process are too inflexible, Federal agencies are encouraged to set time limits appropriate to individual actions (consistent with the time intervals required by § 1506.10). When multiple agencies are involved the reference to agency below means lead agency.

(a) The agency shall set time limits if an applicant for the proposed action requests them: *Provided,* That the limits are consistent with the purposes of NEPA and other essential considerations of national policy.

(b) The agency may:

(1) Consider the following factors in determining time limits:

(i) Potential for environmental harm.

(ii) Size of the proposed action.

(iii) State of the art of analytic techniques.

(iv) Degree of public need for the proposed action, including the consequences of delay.

(v) Number of persons and agencies affected.

(vi) Degree to which relevant information is known and if not known the time required for obtaining it.

(vii) Degree to which the action is controversial.

(viii) Other time limits imposed on the agency by law, regulations, or executive order.

(2) Set overall time limits or limits for each constituent part of the NEPA process, which may include:

(i) Decision on whether to prepare an environmental impact statement (if not already decided).

(ii) Determination of the scope of the environmental impact statement.

(iii) Preparation of the draft environmental impact statement.

(iv) Review of any comments on the draft environmental impact statement from the public and agencies.

(v) Preparation of the final environmental impact statement.

(vi) Review of any comments on the final environmental impact statement.

(vii) Decision on the action based in part on the environmental impact statement.

(3) Designate a person (such as the project manager or a person in the agency's office with NEPA responsibilities) to expedite the NEPA process.

(c) State or local agencies or members of the public may request a Federal Agency to set time limits.

PART 1502—ENVIRONMENTAL IMPACT STATEMENT

Sec.
1502.1   Purpose.
1502.2   Implementation.
1502.3   Statutory Requirements for Statements.
1502.4   Major Federal Actions Requiring the Preparation of Environmental Impact Statements.
1502.5   Timing.
1502.6   Interdisciplinary Preparation.
1502.7   Page Limits.
1502.8   Writing.
1502.9   Draft, Final, and Supplemental Statements.
1502.10   Recommended Format.
1502.11   Cover Sheet.
1502.12   Summary.
1502.13   Purpose and Need.
1502.14   Alternatives Including the Proposed Action.
1502.15   Affected Environment.
1502.16   Environmental Consequences.
1502.17   List of Preparers.
1502.18   Appendix.
1502.19   Circulation of the Environmental Impact Statement.
1502.20   Tiering.
1502.21   Incorporation by Reference.
1502.22   Incomplete or Unavailable Information.
1502.23   Cost-Benefit Analysis.
1502.24   Methodology and Scientific Accuracy.
1502.25   Environmental Review and Consultation Requirements.

AUTHORITY: NEPA, the Environmental Quality Improvement Act of 1970, as amended (42 U.S.C. 4371 et seq.), Section 309 of the Clean Air Act, as amended (42 U.S.C. 7609), and Executive Order 11514, Protection and Enhancement of Environmental Quality (March 5, 1970, as amended by Executive Order 11991, May 24, 1977).

§ 1502.1   Purpose.

The primary purpose of an environmental impact statement is to serve as an action-forcing device to insure that the policies and goals defined in the Act are infused into the ongoing programs and actions of the Federal Government. It shall provide full and fair discussion of significant environmental impacts and shall inform decisionmakers and the public of the reasonable alternatives which would avoid or minimize adverse impacts or enhance the quality of the human environment. Agencies shall focus on significant environmental issues and alternatives and shall reduce paperwork and the accumulation of extraneous background data. Statements shall be concise, clear, and to the point, and shall be supported by evidence that the agency has made the necessary environmental analyses. An environmental impact statement is more than a disclosure document. It shall be used by Federal officials in conjunction with other relevant material to plan actions and make decisions.

§ 1502.2   Implementation.

To achieve the purposes set forth in § 1502.1 agencies shall prepare environmental impact statements in the following manner:

(a) Environmental impact statements shall be analytic rather than encyclopedic.

(b) Impacts shall be discussed in proportion to their significance. There shall be only brief discussion of other than significant issues. As in a finding of no significant impact, there should be only enough discussion to show why more study is not warranted.

(c) Environmental impact statements shall be kept concise and shall be no longer than absolutely necessary to comply with NEPA and with these regulations. Length should vary first with potential environmental problems and then with project size.

(d) Environmental impact statements shall state how alternatives considered in it and decisions based on it will or will not achieve the requirements of sections 101 and 102(1) of the Act and other environmental laws and policies.

(e) The range of alternatives discussed in environmental impact statements shall encompass those to be considered by the ultimate agency decisionmaker.

(f) Agencies shall not commit resources prejudicing selection of alternatives before making a final decision (§ 1506.1).

(g) Environmental impact statements shall serve as the means of assessing the environmental impact of proposed agency actions, rather than justifying decisions already made.

§ 1502.3 Statutory requirements for statements.

As required by sec. 102(2)(C) of NEPA environmental impact statements (§ 1508.11) are to be included in every recommendation or report
 On proposals (§ 1508.23)
 For legislation and (§ 1508.17)
 Other major Federal actions (§ 1508.18)
 Significantly (§ 1508.27)
 Affecting (§§ 1508.3, 1508.8)
The quality of the human environment (§ 1508.14).

§ 1502.4 Major Federal actions requiring the preparation of environmental impact statements.

(a) Agencies shall make sure the proposal which is the subject of an environmental impact statement is properly defined. Agencies shall use the criteria for scope (§ 1508.25) to determine which proposal(s) shall be the subject of a particular statement. Proposals or parts of proposals which are related to each other closely enough to be, in effect, a single course of action shall be evaluated in a single impact statement.

(b) Environmental impact statements may be prepared, and are sometimes required, for broad Federal actions such as the adoption of new agency programs or regulations (§ 1508.18). Agencies shall prepare statements on broad actions so that they are relevant to policy and are timed to coincide with meaningful points in agency planning and decisionmaking.

(c) When preparing statements on broad actions (including proposals by more than one agency), agencies may find it useful to evaluate the proposal(s) in one of the following ways:

(1) Geographically, including actions occurring in the same general location, such as body of water, region, or metropolitan area.

(2) Generically, including actions which have relevant similarities, such as common timing, impacts, alternatives, methods of implementation, media, or subject matter.

(3) By stage of technological development including federal or federally assisted research, development or demonstration programs for new technologies which, if applied, could significantly affect the quality of the human environment. Statements shall be prepared on such programs and shall be available before the program has reached a stage of investment or commitment to implementation likely to determine subsequent development or restrict later alternatives.

(d) Agencies shall as appropriate employ scoping (§ 1501.7), tiering (§ 1502.20), and other methods listed in §§ 1500.4 and 1500.5 to relate broad and narrow actions and to avoid duplication and delay.

§ 1502.5 Timing.

An agency shall commence preparation of an environmental impact statement as close as possible to the time the agency is developing or is presented with a proposal (§ 1508.23) so that preparation can be completed in time for the final statement to be included in any recommendation or report on the proposal. The statement shall be prepared early enough so that it can serve practically as an important contribution to the decisionmaking process and will not be used to rationalize or justify decisions already made (§§ 1500.2(c), 1501.2, and 1502.2). For instance:

(a) For projects directly undertaken by Federal agencies the environmental impact statement shall be prepared at the feasibility analysis (go-no go) stage and may be supplemented at a later stage if necessary.

(b) For applications to the agency appropriate environmental assessments or statements shall be commenced no later than immediately after the application is received. Federal agencies are encouraged to begin preparation of such assessments or statements earlier, preferably jointly with applicable State or local agencies.

(c) For adjudication, the final environmental impact statement shall normally precede the final staff recommendation and that portion of the public hearing related to the impact study. In appropriate circumstances the statement may follow preliminary hearings designed to gather information for use in the statements.

(d) For informal rulemaking the draft environmental impact statement shall normally accompany the proposed rule.

§ 1502.6 Interdisciplinary preparation.

Environmental impact statements shall be prepared using an inter-disciplinary approach which will insure the integrated use of the natural and social sciences and the environmental design arts (section 102(2)(A) of the Act). The disciplines of the preparers shall be appropriate to the scope and issues identified in the scoping process (§ 1501.7).

§ 1502.7 Page limits.

The text of final environmental impact statements (e.g., paragraphs (d) through (g) of § 1502.10) shall normally be less than 150 pages and for proposals of unusual scope or complexity shall normally be less than 300 pages.

§ 1502.8 Writing.

Environmental impact statements shall be written in plain language and may use appropriate graphics so that decisionmakers and the public can readily understand them. Agencies should employ writers of clear prose or editors to write, review, or edit statements, which will be based upon the analysis and supporting data from the natural and social sciences and the environmental design arts.

§ 1502.9 Draft, final, and supplemental statements.

Except for proposals for legislation as provided in § 1506.8 environmental impact statements shall be prepared in two stages and may be supplemented.

(a) Draft environmental impact statements shall be prepared in accordance with the scope decided upon in the scoping process. The lead agency shall work with the cooperating agencies and shall obtain comments as required in Part 1503 of this chapter. The draft statement must fulfill and satisfy to the fullest extent possible the requirements established for final statements in section 102(2)(C) of the Act. If a draft statement is so inadequate as to preclude meaningful analysis, the agency shall prepare and circulate a revised draft of the appropriate portion. The agency shall make every effort to disclose and discuss at appropriate points in the draft statement all major points of view on the environmental impacts of the alternatives including the proposed action.

(b) Final environmental impact statements shall respond to comments as required in Part 1503 of this chapter. The agency shall discuss at appropriate points in the final statement any responsible opposing view which was not adequately discussed in the draft statement and shall indicate the agency's response to the issues raised.

(c) Agencies:

(1) Shall prepare supplements to either draft or final environmental impact statements if:

(i) The agency makes substantial changes in the proposed action that are relevant to environmental concerns; or

(ii) There are significant new circumstances or information relevant to environmental concerns and bearing on the proposed action or its impacts.

(2) May also prepare supplements when the agency determines that the purposes of the Act will be furthered by doing so.

(3) Shall adopt procedures for introducing a supplement into its formal administrative record, if such a record exists.

(4) Shall prepare, circulate, and file a supplement to a statement in

the same fashion (exclusive of scoping) as a draft and final statement unless alternative procedures are approved by the Council.

§ 1502.10 Recommended format.

Agencies shall use a format for environmental impact statements which will encourage good analysis and clear presentation of the alternatives including the proposed action. The following standard format for environmental impact statements should be followed unless the agency determines that there is a compelling reason to do otherwise:
(a) Cover sheet.
(b) Summary.
(c) Table of Contents.
(d) Purpose of and Need for Action.
(e) Alternatives Including Proposed Action (secs. 102(2)(C)(iii) and 102(2)(E) of the Act).
(f) Affected Environment.
(g) Environmental Consequences (especially sections 102(2)(C) (i), (ii), (iv), and (v) of the Act).
(h) List of Preparers.
(i) List of Agencies, Organizations, and Persons to Whom Copies of the Statement Are Sent.
(j) Index.
(k) Appendices (if any).

If a different format is used, it shall include paragraphs (a), (b), (c), (h), (i), and (j), of this section and shall include the substance of paragraphs (d), (e), (f), (g), and (k) of this section, as further described in §§ 1502.11-1502.18, in any appropriate format.

§ 1502.11 Cover sheet.

The cover sheet shall not exceed one page. It shall include:
(a) A list of the responsible agencies including the lead agency and any cooperating agencies.
(b) The title of the proposed action that is the subject of the statement (and if appropriate the titles of related cooperating agency actions), together with the State(s) and county(ies) (or other jurisdiction if applicable) where the action is located.
(c) The name, address, and telephone number of the person at the agency who can supply further information.
(d) A designation of the statement as a draft, final, or draft or final supplement.
(e) A one paragraph abstract of the statement.
(f) The date by which comments must be received (computed in cooperation with EPA under § 1506.10).

The information required by this section may be entered on Standard Form 424 (in items 4, 6, 7, 10, and 18).

§ 1502.12 Summary.

Each environmental impact statement shall contain a summary which adequately and accurately summarizes the statement. The summary shall stress the major conclusions, areas of controversy (including issues raised by agencies and the public), and the issues to be resolved (including the choice among alternatives). The summary will normally not exceed 15 pages.

§ 1502.13 Purpose and need.

The statement shall briefly specify the underlying purpose and need to which the agency is responding in proposing the alternatives including the proposed action.

§ 1502.14 Alternatives including the proposed action.

This section is the heart of the environmental impact statement. Based on the information and analysis presented in the sections on the Affected Environment (§ 1502.15) and the Environmental Consequences (§ 1502.16), it should present the environmental impacts of the proposal and the alternatives in comparative form, thus sharply defining the issues and providing a clear basis for choice among options by the decisionmaker and the public. In this section agencies shall:
(a) Rigorously explore and objectively evaluate all reasonable alternatives, and for alternatives which were eliminated from detailed study, briefly discuss the reasons for their having been eliminated.
(b) Devote substantial treatment to each alternative considered in detail including the proposed action

so that reviewers may evaluate their comparative merits.

(c) Include reasonable alternatives not within the jurisdiction of the lead agency.

(d) Include the alternative of no action.

(e) Identify the agency's preferred alternative or alternatives, if one or more exists, in the draft statement and identify such alternative in the final statement unless another law prohibits the expression of such a preference.

(f) Include appropriate mitigation measures not already included in the proposed action or alternatives.

§ 1502.15 Affected environment.

The environmental impact statement shall succinctly describe the environment of the area(s) to be affected or created by the alternatives under consideration. The descriptions shall be no longer than is necessary to understand the effects of the alternatives. Data and analyses in a statement shall be commensurate with the importance of the impact, with less important material summarized, consolidated, or simply referenced. Agencies shall avoid useless bulk in statements and shall concentrate effort and attention on important issues. Verbose descriptions of the affected environment are themselves no measure of the adequacy of an environmental impact statement.

§ 1502.16 Environmental consequences.

This section forms the scientific and analytic basis for the comparisons under § 1502.14. It shall consolidate the discussions of those elements required by secs. 102(2)(C) (i), (ii), (iv), and (v) of NEPA which are within the scope of the statement and as much of sec. 102(2)(C)(iii) as is necessary to support the comparisons. The discussion will include the environmental impacts of the alternatives including the proposed action, any adverse environmental effects which cannot be avoided should the proposal be implemented, the relationship between short-term uses of man's environment and the maintenance and enhancement of long-term productivity, and any irreversible or irretrievable commitments of resources which would be involved in the proposal should it be implemented. This section should not duplicate discussions in § 1502.14. It shall include discussions of:

(a) Direct effects and their significance (§ 1508.8).

(b) Indirect effects and their significance (§ 1508.8).

(c) Possible conflicts between the proposed action and the objectives of Federal, regional, State, and local (and in the case of a reservation, Indian tribe) land use plans, policies and controls for the area concerned. (See § 1506.2(d).)

(d) The environmental effects of alternatives including the proposed action. The comparisons under § 1502.14 will be based on this discussion.

(e) Energy requirements and conservation potential of various alternatives and mitigation measures.

(f) Natural or depletable resource requirements and conservation potential of various alternatives and mitigation measures.

(g) Urban quality, historic and cultural resources, and the design of the built environment, including the reuse and conservation potential of various alternatives and mitigation measures.

(h) Means to mitigate adverse environmental impacts (if not fully covered under § 1502.14(f)).

§ 1502.17 List of preparers.

The environmental impact statement shall list the names, together with their qualifications (expertise, experience, professional disciplines), of the persons who were primarily responsible for preparing the environmental impact statement or significant background papers, including basic components of the statement (§§ 1502.6 and 1502.8). Where possible the persons who are responsible for a particular analysis, including analyses in background papers, shall be identified. Normally the list will not exceed two pages.

§ 1502.18 Appendix.

If an agency prepares an appendix to an environmental impact statement the appendix shall:

(a) Consist of material prepared in connection with an environmental impact statement (as distinct from material which is not so prepared and which is incorporated by reference (§ 1502.21)).

(b) Normally consist of material which substantiates any analysis fundamental to the impact statement.

(c) Normally be analytic and relevant to the decision to be made.

(d) Be circulated with the environmental impact statement or be readily available on request.

§ 1502.19 Circulation of the environmental impact statement.

Agencies shall circulate the entire draft and final environmental impact statements except for certain appendices as provided in § 1502.18(d) and unchanged statements as provided in § 1503.4(c). However, if the statement is unusually long, the agency may circulate the summary instead, except that the entire statement shall be furnished to:

(a) Any Federal agency which has jurisdiction by law or special expertise with respect to any environmental impact involved and any appropriate Federal, State or local agency authorized to develop and enforce environmental standards.

(b) The applicant, if any.

(c) Any person, organization, or agency requesting the entire environmental impact statement.

(d) In the case of a final environmental impact statement any person, organization, or agency which submitted substantive comments on the draft.

If the agency circulates the summary and thereafter receives a timely request for the entire statement and for additional time to comment, the time for that requestor only shall be extended by at least 15 days beyond the minimum period.

§ 1502.20 Tiering.

Agencies are encouraged to tier their environmental impact statements to eliminate repetitive discussions of the same issues and to focus on the actual issues ripe for decision at each level of environmental review (§ 1508.28). Whenever a broad environmental impact statement has been prepared (such as a program or policy statement) and a subsequent statement or environmental assessment is then prepared on an action included within the entire program or policy (such as a site specific action) the subsequent statement or environmental assessment need only summarize the issues discussed in the broader statement and incorporate discussions from the broader statement by reference and shall concentrate on the issues specific to the subsequent action. The subsequent document shall state where the earlier document is available. Tiering may also be appropriate for different stages of actions. (Sec. 1508.28).

§ 1502.21 Incorporation by reference.

Agencies shall incorporate material into an environmental impact statement by reference when the effect will be to cut down on bulk without impeding agency and public review of the action. The incorporated material shall be cited in the statement and its content briefly described. No material may be incorporated by reference unless it is reasonably available for inspection by potentially interested persons within the time allowed for comment. Material based on proprietary data which is itself not available for review and comment shall not be incorporated by reference.

§ 1502.22 Incomplete or unavailable information.

When an agency is evaluating significant adverse effects on the human environment in an environmental impact statement and there are gaps in relevant information or scientific uncertainty, the agency shall always make clear that such information is lacking or that uncertainty exists.

(a) If the information relevant to adverse impacts is essential to a reasoned choice among alternatives and is not known and the overall costs of obtaining it are not exorbitant, the agency shall include the information in the environmental impact statement.

(b) If (1) the information relevant to adverse impacts is essential to a reasoned choice among alternatives and is not known and the overall costs of obtaining it are exorbitant or (2) the information relevant to adverse impacts is important to the decision and the means to obtain it are not known (e.g., the means for obtaining it are beyond the state of the art) the agency shall weigh the need for the action against the risk and severity of possible adverse impacts were the action to proceed in the face of uncertainty. If the agency proceeds, it shall include a worst case analysis and an indication of the probability or improbability of its occurrence.

§ 1502.23 Cost-benefit analysis.

If a cost-benefit analysis relevant to the choice among environmentally different alternatives is being considered for the proposed action, it shall be incorporated by reference or appended to the statement as an aid in evaluating the environmental consequences. To assess the adequacy of compliance with sec. 102(2)(B) of the Act the statement shall, when a cost-benefit analysis is prepared, discuss the relationship between that analysis and any analyses of unquantified environmental impacts, values, and amenities. For purposes of complying with the Act, the weighing of the merits and drawbacks of the various alternatives need not be displayed in a monetary cost-benefit analysis and should not be when there are important qualitative considerations. In any event, an environmental impact statement should at least indicate those considerations, including factors not related to environmental quality, which are likely to be relevant and important to a decision.

§ 1502.24 Methodology and scientific accuracy.

Agencies shall insure the professional integrity, including scientific integrity, of the discussions and analyses in environmental impact statements. They shall identify any methodologies used and shall make explicit reference by footnote to the scientific and other sources relied upon for conclusions in the statement. An agency may place discussion of methodology in an appendix.

§ 1502.25 Environmental review and consultation requirements.

(a) To the fullest extent possible, agencies shall prepare draft environmental impact statements concurrently with and integrated with environmental impact analyses and related surveys and studies required by the Fish and Wildlife Coordination Act (16 U.S.C. Sec. 661 et seq.), the National Historic Preservation Act of 1966 (16 U.S.C. Sec. 470 et seq.), the Endangered Species Act of 1973 (16 U.S.C. Sec. 1531 et seq.), and other environmental review laws and executive orders.

(b) The draft environmental impact statement shall list all Federal permits, licenses, and other entitlements which must be obtained in implementing the proposal. If it is uncertain whether a Federal permit, license, or other entitlement is necessary, the draft environmental impact statement shall so indicate.

PART 1503—COMMENTING

Sec.
1503.1 Inviting Comments.
1503.2 Duty to Comment.
1503.3 Specificity of Comments.
1503.4 Response to Comments.

AUTHORITY: NEPA, the Environmental Quality Improvement Act of 1970, as amended (42 U.S.C. 4371 et seq.), Section 309 of the Clean Air Act, as amended (42 U.S.C. 7609), and Executive Order 11514, Protection and Enhancement of Environmental Quality (March 5, 1970, as amended by Executive Order 11991, May 24, 1977).

§ 1503.1 Inviting comments.

(a) After preparing a draft environmental impact statement and before preparing a final environmental impact statement the agency shall:

(1) Obtain the comments of any Federal agency which has jurisdic-

tion by law or special expertise with respect to any environmental impact involved or which is authorized to develop and enforce environmental standards.

(2) Request the comments of:

(i) Appropriate State and local agencies which are authorized to develop and enforce environmental standards;

(ii) Indian tribes, when the effects may be on a reservation; and

(iii) Any agency which has requested that it receive statements on actions of the kind proposed.

Office of Management and Budget Circular A-95 (Revised), through its system of clearinghouses, provides a means of securing the views of State and local environmental agencies. The clearinghouses may be used, by mutual agreement of the lead agency and the clearinghouse, for securing State and local reviews of the draft environmental impact statements.

(3) Request comments from the applicant, if any.

(4) Request comments from the public, affirmatively soliciting comments from those persons or organizations who may be interested or affected.

(b) An agency may request comments on a final environmental impact statement before the decision is finally made. In any case other agencies or persons may make comments before the final decision unless a different time is provided under § 1506.10.

§ 1503.2 Duty to comment.

Federal agencies with jurisdiction by law or special expertise with respect to any environmental impact involved and agencies which are authorized to develop and enforce environmental standards shall comment on statements within their jurisdiction, expertise, or authority. Agencies shall comment within the time period specified for comment in § 1506.10. A Federal agency may reply that it has no comment. If a cooperating agency is satisfied that its views are adequately reflected in the environmental impact statement, it should reply that it has no comment.

§ 1503.3 Specificity of comments.

(a) Comments on an environmental impact statement or on a proposed action shall be as specific as possible and may address either the adequacy of the statement or the merits of the alternatives discussed or both.

(b) When a commenting agency criticizes a lead agency's predictive methodology, the commenting agency should describe the alternative methodology which it prefers and why.

(c) A cooperating agency shall specify in its comments whether it needs additional information to fulfill other applicable environmental reviews or consultation requirements and what information it needs. In particular, it shall specify any additional information it needs to comment adequately on the draft statement's analysis of significant site-specific effects associated with the granting or approving by that cooperating agency of necessary Federal permits, licenses, or entitlements.

(d) When a cooperating agency with jurisdiction by law objects to or expresses reservations about the proposal on grounds of environmental impacts, the agency expressing the objection or reservation shall specify the mitigation measures it considers necessary to allow the agency to grant or approve applicable permit, license, or related requirements or concurrences.

§ 1503.4 Response to comments.

(a) An agency preparing a final environmental impact statement shall assess and consider comments both individually and collectively, and shall respond by one or more of the means listed below, stating its response in the final statement. Possible responses are to:

(1) Modify alternatives including the proposed action.

(2) Develop and evaluate alternatives not previously given serious consideration by the agency.

(3) Supplement, improve, or modify its analyses.

(4) Make factual corrections.

(5) Explain why the comments do

not warrant further agency response, citing the sources, authorities, or reasons which support the agency's position and, if appropriate, indicate those circumstances which would trigger agency reappraisal or further response.

(b) All substantive comments received on the draft statement (or summaries thereof where the response has been exceptionally voluminous), should be attached to the final statement whether or not the comment is thought to merit individual discussion by the agency in the text of the statement.

(c) If changes in response to comments are minor and are confined to the responses described in paragraphs (a) (4) and (5) of this section, agencies may write them on errata sheets and attach them to the statement instead of rewriting the draft statement. In such cases only the comments, the responses, and the changes and not the final statement need be circulated (§ 1502.19). The entire document with a new cover sheet shall be filed as the final statement (§ 1506.9).

PART 1504—PREDECISION REFERRALS TO THE COUNCIL OF PROPOSED FEDERAL ACTIONS DETERMINED TO BE ENVIRONMENTALLY UNSATISFACTORY

Sec.
1504.1 Purpose.
1504.2 Criteria for Referral.
1504.3 Procedure for Referrals and Response.

AUTHORITY: NEPA, the Environmental Quality Improvement Act of 1970, as amended (42 U.S.C. 4371 et seq.), Section 309 of the Clean Air Act, as amended (42 U.S.C. 7609), and Executive Order 11514, Protection and Enhancement of Environmental Quality (March 5, 1970, as amended by Executive Order 11991, May 24, 1977).

§ 1504.1 Purpose.

(a) This part establishes procedures for referring to the Council Federal interagency disagreements concerning proposed major Federal actions that might cause unsatisfactory environmental effects. It provides means for early resolution of such disagreements.

(b) Under section 309 of the Clean Air Act (42 U.S.C. 7609), the Administrator of the Environmental Protection Agency is directed to review and comment publicly on the environmental impacts of Federal activities, including actions for which environmental impact statements are prepared. If after this review the Administrator determines that the matter is "unsatisfactory from the standpoint of public health or welfare or environmental quality," section 309 directs that the matter be referred to the Council (hereafter "environmental referrals").

(c) Under section 102(2)(C) of the Act other Federal agencies may make similar reviews of environmental impact statements, including judgments on the acceptability of anticipated environmental impacts. These reviews must be made available to the President, the Council and the public.

§ 1504.2 Criteria for referral.

Environmental referrals should be made to the Council only after concerted, timely (as early as possible in the process), but unsuccessful attempts to resolve differences with the lead agency. In determining what environmental objections to the matter are appropriate to refer to the Council, an agency should weigh potential adverse environmental impacts, considering:

(a) Possible violation of national environmental standards or policies.

(b) Severity.

(c) Geographical scope.

(d) Duration.

(e) Importance as precedents.

(f) Availability of environmentally preferable alternatives.

§ 1504.3 Procedure for referrals and response.

(a) A Federal agency making the referral to the Council shall:

(1) Advise the lead agency at the earliest possible time that it intends to refer a matter to the Council unless a satisfactory agreement is reached.

(2) Include such advice in the referring agency's comments on the draft environmental impact statement, except when the statement does not contain adequate information to permit an assessment of the matter's environmental acceptability.

(3) Identify any essential information that is lacking and request that it be made available at the earliest possible time.

(4) Send copies of such advice to the Council.

(b) The referring agency shall deliver its referral to the Council not later than twenty-five (25) days after the final environmental impact statement has been made available to the Environmental Protection Agency, commenting agencies, and the public. Except when an extension of this period has been granted by the lead agency, the Council will not accept a referral after that date.

(c) The referral shall consist of:

(1) A copy of the letter signed by the head of the referring agency and delivered to the lead agency informing the lead agency of the referral and the reasons for it, and requesting that no action be taken to implement the matter until the Council acts upon the referral. The letter shall include a copy of the statement referred to in (c)(2) below.

(2) A statement supported by factual evidence leading to the conclusion that the matter is unsatisfactory from the standpoint of public health or welfare or environmental quality. The statement shall:

(i) Identify any material facts in controversy and incorporate (by reference if appropriate) agreed upon facts,

(ii) Identify any existing environmental requirements or policies which would be violated by the matter,

(iii) Present the reasons why the referring agency believes the matter is environmentally unsatisfactory,

(iv) Contain a finding by the agency whether the issue raised is of national importance because of the threat to national environmental resources or policies or for some other reason,

(v) Review the steps taken by the referring agency to bring its concerns to the attention of the lead agency at the earliest possible time, and

(vi) Give the referring agency's recommendations as to what mitigation alternative, further study, or other course of action (including abandonment of the matter) are necessary to remedy the situation.

(d) Not later than twenty-five (25) days after the referral to the Council the lead agency may deliver a response to the Council and the referring agency. If the lead agency requests more time and gives assurance that the matter will not go forward in the interim, the Council may grant an extension. The response shall:

(1) Address fully the issues raised in the referral.

(2) Be supported by evidence.

(3) Give the lead agency's response to the referring agency's recommendations.

(e) Interested persons (including the applicant) may deliver their views in writing to the Council. Views in support of the referral should be delivered not later than the referral. Views in support of the response shall be delivered not later than the response.

(f) Not later than twenty-five (25) days after receipt of both the referral and any response or upon being informed that there will be no response (unless the lead agency agrees to a longer time), the Council may take one or more of the following actions:

(1) Conclude that the process of referral and response has successfully resolved the problem.

(2) Initiate discussions with the agencies with the objective of mediation with referring and lead agencies.

(3) Hold public meetings or hearings to obtain additional views and information.

(4) Determine that the issue is not one of national importance and request the referring and lead agencies to pursue their decision process.

(5) Determine that the issue should be further negotiated by the referring and lead agencies and is not appropriate for Council consideration until one or more heads of

agencies report to the Council that the agencies' disagreements are irreconcilable.

(6) Publish its findings and recommendations (including where appropriate a finding that the submitted evidence does not support the position of an agency).

(7) When appropriate, submit the referral and the response together with the Council's recommendation to the President for action.

(g) The Council shall take no longer than 60 days to complete the actions specified in paragraph (f)(2), (3), or (5) of this section.

(h) When the referral involves an action required by statute to be determined on the record after opportunity for agency hearing, the referral shall be conducted in a manner consistent with 5 U.S.C. 557(d) (Administrative Procedure Act).

PART 1505—NEPA AND AGENCY DECISIONMAKING

Sec.
1505.1 Agency decisionmaking procedures.
1505.2 Record of decision in cases requiring environmental impact statements.
1505.3 Implementing the decision.

AUTHORITY: NEPA, the Environmental Quality Improvement Act of 1970, as amended (42 U.S.C. 4371 et seq.), Section 309 of the Clean Air Act, as amended (42 U.S.C. 7609), and Executive Order 11514, Protection and Enhancement of Environmental Quality (March 5, 1970, as amended by Executive Order 11991, May 24, 1977).

§ 1505.1 **Agency decisionmaking procedures.**

Agencies shall adopt procedures (§ 1507.3) to ensure that decisions are made in accordance with the policies and purposes of the Act. Such procedures shall include but not be limited to:

(a) Implementing procedures under section 102(2) to achieve the requirements of sections 101 and 102(1).

(b) Designating the major decision points for the agency's principal programs likely to have a significant effect on the human environment and assuring that the NEPA process corresponds with them.

(c) Requiring that relevant environmental documents, comments, and responses be part of the record in formal rulemaking or adjudicatory proceedings.

(d) Requiring that relevant environmental documents, comments, and responses accompany the proposal through existing agency review processes so that agency officials use the statement in making decisions.

(e) Requiring that the alternatives considered by the decisionmaker are encompassed by the range of alternatives discussed in the relevant environmental documents and that the decisionmaker consider the alternatives described in the environmental impact statement. If another decision document accompanies the relevant environmental documents to the decisionmaker, agencies are encouraged to make available to the public before the decision is made any part of that document that relates to the comparison of alternatives.

§ 1505.2 **Record of decision in cases requiring environmental impact statements.**

At the time of its decision (§ 1506.10) or, if appropriate, its recommendation to Congress, each agency shall prepare a concise public record of decision. The record, which may be integrated into any other record prepared by the agency, including that required by OMB Circular A-95 (Revised), part I, sections 6 (c) and (d), and part II, section 5(b)(4), shall:

(a) State what the decision was.

(b) Identify all alternatives considered by the agency in reaching its decision, specifying the alternative or alternatives which were considered to be environmentally preferable. An agency may discuss preferences among alternatives based on relevant factors including economic and technical considerations and agency statutory missions. An agency shall identify and discuss all such factors including any essential considerations of national policy which were balanced by the agency in making its decision and state how

APPENDIXES / 117

those considerations entered into its decision.

(c) State whether all practicable means to avoid or minimize environmental harm from the alternative selected have been adopted, and if not, why they were not. A monitoring and enforcement program shall be adopted and summarized where applicable for any mitigation.

§ 1505.3 Implementing the decision.

Agencies may provide for monitoring to assure that their decisions are carried out and should do so in important cases. Mitigation (§ 1505.2(c)) and other conditions established in the environmental impact statement or during its review and committed as part of the decision shall be implemented by the lead agency or other appropriate consenting agency. The lead agency shall:

(a) Include appropriate conditions in grants, permits or other approvals.

(b) Condition funding of actions on mitigation.

(c) Upon request, inform cooperating or commenting agencies on progress in carrying out mitigation measures which they have proposed and which were adopted by the agency making the decision.

(d) Upon request, make available to the public the results of relevant monitoring.

PART 1506—OTHER REQUIREMENTS OF NEPA

Sec.
1506.1 Limitations on actions during NEPA process.
1506.2 Elimination of duplication with State and local procedures.
1506.3 Adoption.
1506.4 Combining documents.
1506.5 Agency responsibility.
1506.6 Public involvement.
1506.7 Further guidance.
1506.8 Proposals for legislation.
1506.9 Filing requirements.
1506.10 Timing of agency action.
1506.11 Emergencies.
1506.12 Effective date.

AUTHORITY: NEPA, the Environmental Quality Improvement Act of 1970, as amended (42 U.S.C. 4371 et seq.), Section 309 of the Clean Air Act, as amended (42 U.S.C. 7609), and Executive Order 11514, Protection and Enhancement of Environmental Quality (March 5, 1970, as amended by Executive Order 11991, May 24, 1977).

§ 1506.1 Limitations on actions during NEPA process.

(a) Until an agency issues a record of decision as provided in § 1505.2 (except as provided in paragraph (c) of this section), no action concerning the proposal shall be taken which would:

(1) Have an adverse environmental impact; or

(2) Limit the choice of reasonable alternatives.

(b) If any agency is considering an application from a non-Federal entity, and is aware that the applicant is about to take an action within the agency's jurisdiction that would meet either of the criteria in paragraph (a) of this section, then the agency shall promptly notify the applicant that the agency will take appropriate action to insure that the objectives and procedures of NEPA are achieved.

(c) While work on a required program environmental impact statement is in progress and the action is not covered by an existing program statement, agencies shall not undertake in the interim any major Federal action covered by the program which may significantly affect the quality of the human environment unless such action:

(1) Is justified independently of the program;

(2) Is itself accompanied by an adequate environmental impact statement; and

(3) Will not prejudice the ultimate decision on the program. Interim action prejudices the ultimate decision on the program when it tends to determine subsequent development or limit alternatives.

(d) This section does not preclude development by applicants of plans or designs or performance of other work necessary to support an application for Federal, State or local permits or assistance. Nothing in this section shall preclude Rural Electrification Administration approval of minimal expenditures not affecting the environment (*e.g.* long leadtime equipment and purchase options) made by non-governmental entities seeking loan guarantees from the Administration.

§ 1506.2 Elimination of duplication with State and local procedures.

(a) Agencies authorized by law to cooperate with State agencies of statewide jurisdiction pursuant to section 102(2)(D) of the Act may do so.

(b) Agencies shall cooperate with State and local agencies to the fullest extent possible to reduce duplication between NEPA and State and local requirements, unless the agencies are specifically barred from doing so by some other law. Except for cases covered by paragraph (a) of this section, such cooperation shall to the fullest extent possible include:

(1) Joint planning processes.

(2) Joint environmental research and studies.

(3) Joint public hearings (except where otherwise provided by statute).

(4) Joint environmental assessments.

(c) Agencies shall cooperate with State and local agencies to the fullest extent possible to reduce duplication between NEPA and comparable State and local requirements, unless the agencies are specifically barred from doing so by some other law. Except for cases covered by paragraph (a) of this section, such cooperation shall to the fullest extent possible include joint environmental impact statements. In such cases one or more Federal agencies and one or more State or local agencies shall be joint lead agencies. Where State laws or local ordinances have environmental impact statement requirements in addition to but not in conflict with those in NEPA, Federal agencies shall cooperate in fulfilling these requirements as well as those of Federal laws so that one document will comply with all applicable laws.

(d) To better integrate environmental impact statements into State or local planning processes, statements shall discuss any inconsistency of a proposed action with any approved State or local plan and laws (whether or not federally sanctioned). Where an inconsistency exists, the statement should describe the extent to which the agency would reconcile its proposed action with the plan or law.

§ 1506.3 Adoption.

(a) An agency may adopt a Federal draft or final environmental impact statement or portion thereof provided that the statement or portion thereof meets the standards for an adequate statement under these regulations.

(b) If the actions covered by the original environmental impact statement and the proposed action are substantially the same, the agency adopting another agency's statement is not required to recirculate it except as a final statement. Otherwise the adopting agency shall treat the statement as a draft and recirculate it (except as provided in paragraph (c) of this section).

(c) A cooperating agency may adopt without recirculating the environmental impact statement of a lead agency when, after an independent review of the statement, the cooperating agency concludes that its comments and suggestions have been satisfied.

(d) When an agency adopts a statement which is not final within the agency that prepared it, or when the action it assesses is the subject of a referral under part 1504, or when the statement's adequacy is the subject of a judicial action which is not final, the agency shall so specify.

§ 1506.4 Combining documents.

Any environmental document in compliance with NEPA may be combined with any other agency document to reduce duplication and paperwork.

§ 1506.5 Agency responsibility.

(a) *Information.* If an agency requires an applicant to submit environmental information for possible use by the agency in preparing an environmental impact statement, then the agency should assist the applicant by outlining the types of information required. The agency shall independently evaluate the information submitted and shall be responsible for its accuracy. If the agency chooses to use the information submitted by the applicant in the environmental impact statement, either directly or by reference, then the names of the persons responsible for the independent evaluation shall be included in the list of preparers (§ 1502.17). It is the intent of this subparagraph that acceptable work not be redone, but that it be verified by the agency.

(b) *Environmental assessments.* If an agency permits an applicant to prepare an environmental assessment, the agency, besides fulfilling the requirements of paragraph (a) of this section, shall make its own evaluation of the environmental issues and take responsibility for the scope and content of the environmental assessment.

(c) *Environmental impact statements.* Except as provided in §§ 1506.2 and 1506.3 any environmental impact statement prepared pursuant to the requirements of NEPA shall be prepared directly by or by a contractor selected by the lead agency or where appropriate under § 1501.6(b), a cooperating agency. It is the intent of these regulations that the contractor be chosen solely by the lead agency, or by the lead agency in cooperation with cooperating agencies, or where appropriate by a cooperating agency to avoid any conflict of interest. Contractors shall execute a disclosure statement prepared by the lead agency, or where appropriate the cooperating agency, specifying that they have no financial or other interest in the outcome of the project. If the document is prepared by contract, the responsible Federal official shall furnish guidance and participate in the preparation and shall independently evaluate the statement prior to its approval and take responsibility for its scope and contents. Nothing in this section is intended to prohibit any agency from requesting any person to submit information to it or to prohibit any person from submitting information to any agency.

§ 1506.6 Public involvement.

Agencies shall: (a) Make diligent efforts to involve the public in preparing and implementing their NEPA procedures.

(b) Provide public notice of NEPA-related hearings, public meetings, and the availability of environmental documents so as to inform those persons and agencies who may be interested or affected.

(1) In all cases the agency shall mail notice to those who have requested it on an individual action.

(2) In the case of an action with effects of national concern notice shall include publication in the FEDERAL REGISTER and notice by mail to national organizations reasonably expected to be interested in the matter and may include listing in the *102 Monitor.* An agency engaged in rulemaking may provide notice by mail to national organizations who have requested that notice regularly be provided. Agencies shall maintain a list of such organizations.

(3) In the case of an action with effects primarily of local concern the notice may include:

(i) Notice to State and areawide clearinghouses pursuant to OMB Circular A-95 (Revised).

(ii) Notice to Indian tribes when effects may occur on reservations.

(iii) Following the affected State's public notice procedures for comparable actions.

(iv) Publication in local newspapers (in papers of general circulation rather than legal papers).

(v) Notice through other local media.

(vi) Notice to potentially interested community organizations including small business associations.

(vii) Publication in newsletters that may be expected to reach potentially interested persons.

(viii) Direct mailing to owners and occupants of nearby or affected property.

(ix) Posting of notice on and off site in the area where the action is to be located.

(c) Hold or sponsor public hearings or public meetings whenever appropriate or in accordance with statutory requirements applicable to the agency. Criteria shall include whether there is:

(1) Substantial environmental controversy concerning the proposed action or substantial interest in holding the hearing.

(2) A request for a hearing by another agency with jurisdiction over the action supported by reasons why a hearing will be helpful. If a draft environmental impact statement is to be considered at a public hearing, the agency should make the statement available to the public at least 15 days in advance (unless the purpose of the hearing is to provide information for the draft environmental impact statement).

(d) Solicit appropriate information from the public.

(e) Explain in its procedures where interested persons can get information or status reports on environmental impact statements and other elements of the NEPA process.

(f) Make environmental impact statements, the comments received, and any underlying documents available to the public pursuant to the provisions of the Freedom of Information Act (5 U.S.C. 552), without regard to the exclusion for interagency memoranda where such memoranda transmit comments of Federal agencies on the environmental impact of the proposed action. Materials to be made available to the public shall be provided to the public without charge to the extent practicable, or at a fee which is not more than the actual costs of reproducing copies required to be sent to other Federal agencies, including the Council.

§ 1506.7 Further guidance.

The Council may provide further guidance concerning NEPA and its procedures including:

(a) A handbook which the Council may supplement from time to time, which shall in plain language provide guidance and instructions concerning the application of NEPA and these regulations.

(b) Publication of the Council's Memoranda to Heads of Agencies.

(c) In conjunction with the Environmental Protection Agency and the publication of the 102 Monitor, notice of:

(1) Research activities;

(2) Meetings and conferences related to NEPA; and

(3) Successful and innovative procedures used by agencies to implement NEPA.

§ 1506.8 Proposals for legislation.

(a) The NEPA process for proposals for legislation (§ 1508.17) significantly affecting the quality of the human environment shall be integrated with the legislative process of the Congress. A legislative environmental impact statement is the detailed statement required by law to be included in a recommendation or report on a legislative proposal to Congress. A legislative environmental impact statement shall be considered part of the formal transmittal of a legislative proposal to Congress; however, it may be transmitted to Congress up to 30 days later in order to allow time for completion of an accurate statement which can serve as the basis for public and Congressional debate. The statement must be available in time for Congressional hearings and deliberations.

(b) Preparation of a legislative environmental impact statement shall conform to the requirements of these regulations except as follows:

(1) There need not be a scoping process.

(2) The legislative statement shall be prepared in the same manner as a draft statement, but shall be considered the "detailed statement" required by statute; *Provided,* That when any of the following conditions exist both the draft and final environmental impact statement on the legislative proposal shall be prepared and circulated as provided by §§ 1503.1 and 1506.10.

(i) A Congressional Committee with jurisdiction over the proposal has a rule requiring both draft and final environmental impact statements.

(ii) The proposal results from a study process required by statute (such as those required by the Wild and Scenic Rivers Act (16 U.S.C.

1271 et seq.) and the Wilderness Act (16 U.S.C. 1131 et seq.)).

(iii) Legislative approval is sought for Federal or federally assisted construction or other projects which the agency recommends be located at specific geographic locations. For proposals requiring an environmental impact statement for the acquisition of space by the General Services Administration, a draft statement shall accompany the Prospectus or the 11(b) Report of Building Project Surveys to the Congress, and a final statement shall be completed before site acquisition.

(iv) The agency decides to prepare draft and final statements.

(c) Comments on the legislative statement shall be given to the lead agency which shall forward them along with its own responses to the Congressional committees with jurisdiction.

§ 1506.9 Filing requirements.

Environmental impact statements together with comments and responses shall be filed with the Environmental Protection Agency, attention Office of Federal Activities (A-104), 401 M Street SW., Washington, D.C. 20460. Statements shall be filed with EPA no earlier than they are also transmitted to commenting agencies and made available to the public. EPA shall deliver one copy of each statement to the Council, which shall satisfy the requirement of availability to the President. EPA may issue guidelines to agencies to implement its responsibilities under this section and § 1506.10 below.

§ 1506.10 Timing of agency action.

(a) The Environmental Protection Agency shall publish a notice in the FEDERAL REGISTER each week of the environmental impact statements filed during the preceding week. The minimum time periods set forth in this section shall be calculated from the date of publication of this notice.

(b) No decision on the proposed action shall be made or recorded under § 1505.2 by a Federal agency until the later of the following dates:

(1) Ninety (90) days after publication of the notice described above in paragraph (a) of this section for a draft environmental impact statement.

(2) Thirty (30) days after publication of the notice described above in paragraph (a) of this section for a final environmental impact statement.

An exception to the rules on timing may be made in the case of an agency decision which is subject to a formal internal appeal. Some agencies have a formally established appeal process which allows other agencies or the public to take appeals on a decision and make their views known, after publication of the final environmental impact statement. In such cases, where a real opportunity exists to alter the decision, the decision may be made and recorded at the same time the environmental impact statement is published. This means that the period for appeal of the decision and the 30-day period prescribed in paragraph (b)(2) of this section may run concurrently. In such cases the environmental impact statement shall explain the timing and the public's right of appeal. An agency engaged in rulemaking under the Administrative Procedure Act or other statute for the purpose of protecting the public health or safety, may waive the time period in paragraph (b)(2) of this section and publish a decision on the final rule simultaneously with publication of the notice of the availability of the final environmental impact statement as described in paragraph (a) of this section.

(c) If the final environmental impact statement is filed within ninety (90) days after a draft environmental impact statement is filed with the Environmental Protection Agency, the minimum thirty (30) day period and the minimum ninety (90) day period may run concurrently. However, subject to paragraph (d) of this section agencies shall allow not less than 45 days for comments on draft statements.

(d) The lead agency may extend prescribed periods. The Environmental Protection Agency may upon a showing by the lead agency of compelling reasons of national policy reduce the prescribed periods and may upon a showing by any other Federal agency of compelling reasons of national policy also extend prescribed periods, but only after consultation with the lead agency. (Also see § 1507.3(d).) Failure to file timely comments shall not be a sufficient reason for extending a period. If the lead agency does not concur with the extension of time, EPA may not extend it for more than 30 days. When the Environmental Protection Agency reduces or extends any period of time it shall notify the Council.

§ 1506.11 Emergencies.

Where emergency circumstances make it necessary to take an action with significant environmental impact without observing the provisions of these regulations, the Federal agency taking the action should consult with the Council about alternative arrangements. Agencies and the Council will limit such arrangements to actions necessary to control the immediate impacts of the emergency. Other actions remain subject to NEPA review.

§ 1506.12 Effective date.

The effective date of these regulations is July 30, 1979, except that for agencies that administer programs that qualify under sec. 102(2)(D) of the Act or under sec. 104(h) of the Housing and Community Development Act of 1974 an additional four months shall be allowed for the State or local agencies to adopt their implementing procedures.

(a) These regulations shall apply to the fullest extent practicable to ongoing activities and environmental documents begun before the effective date. These regulations do not apply to an environmental impact statement or supplement if the draft statement was filed before the effective date of these regulations. No completed environmental documents need be redone by reason of these regulations. Until these regulations are applicable, the Council's guidelines published in the FEDERAL REGISTER of August 1, 1973, shall continue to be applicable. In cases where these regulations are applicable the guidelines are superseded. However, nothing shall prevent an agency from proceeding under these regulations at an earlier time.

(b) NEPA shall continue to be applicable to actions begun before January 1, 1970, to the fullest extent possible.

PART 1507—AGENCY COMPLIANCE

Sec.
1507.1 Compliance.
1507.2 Agency Capability to Comply.
1507.3 Agency Procedures.

AUTHORITY: NEPA, the Environmental Quality Improvement Act of 1970, as amended (42 U.S.C. 4371 et seq.), Section 309 of the Clean Air Act, as amended (42 U.S.C. 7609), and Executive Order 11514, Protection and Enhancement of Environmental Quality (March 5, 1970, as amended by Executive Order 11991, May 24, 1977).

§ 1507.1 Compliance.

All agencies of the Federal Government shall comply with these regulations. It is the intent of these regulations to allow each agency flexibility in adapting its implementing procedures authorized by § 1507.3 to the requirements of other applicable laws.

§ 1507.2 Agency capability to comply.

Each agency shall be capable (in terms of personnel and other resources) of complying with the requirements enumerated below. Such compliance may include use of other's resources, but the using agency shall itself have sufficient capability to evaluate what others do for it. Agencies shall:

(a) Fulfill the requirements of Sec. 102(2)(A) of the Act to utilize a systematic, interdisciplinary approach which will insure the integrated use of the natural and social sciences and the environmental design arts in planning and in decisionmaking which may have an impact on the human environment. Agencies shall designate a person to be responsible for overall review of agency NEPA compliance.

(b) Identify methods and procedures required by Sec. 102(2)(B) to insure that presently unquantified environmental amenities and values may be given appropriate consideration.

(c) Prepare adequate environmental impact statements pursuant to Sec. 102(2)(C) and comment on statements in the areas where the agency has jurisdiction by law or special expertise or is authorized to develop and enforce environmental standards.

(d) Study, develop, and describe alternatives to recommended courses of action in any proposal which involves unresolved conflicts concerning alternative uses of available resources. This requirement of Sec. 102(2)(E) extends to all such proposals, not just the more limited scope of Sec. 102(2)(C)(iii) where the discussion of alternatives is confined to impact statements.

(e) Comply with the requirements of Sec. 102(2)(H) that the agency initiate and utilize ecological information in the planning and development of resource-oriented projects.

(f) Fulfill the requirements of sections 102(2)(F), 102(2)(G), and 102(2)(I), of the Act and of Executive Order 11514, Protection and Enhancement of Environmental Quality, Sec. 2.

§ 1507.3 Agency procedures.

(a) Not later than eight months after publication of these regulations as finally adopted in the FEDERAL REGISTER, or five months after the establishment of an agency, whichever shall come later, each agency shall as necessary adopt procedures to supplement these regulations. When the agency is a department, major subunits are encouraged (with the consent of the department) to adopt their own procedures. Such procedures shall not paraphrase these regulations. They shall confine themselves to implementing procedures. Each agency shall consult with the Council while developing its procedures and before publishing them in the FEDERAL REGISTER for comment. Agencies with similar programs should consult with each other and the Council to coordinate their procedures, especially for programs requesting similar information from applicants. The procedures shall be adopted only after an opportunity for public review and after review by the Council for conformity with the Act and these regulations. The Council shall complete its review within 30 days. Once in effect they shall be filed with the Council and made readily available to the public. Agencies are encouraged to publish explanatory guidance for these regulations and their own procedures. Agencies shall continue to review their policies and procedures and in consultation with the Council to revise them as necessary to ensure full compliance with the purposes and provisions of the Act.

(b) Agency procedures shall comply with these regulations except where compliance would be inconsistent with statutory requirements and shall include:

(1) Those procedures required by §§ 1501.2(d), 1502.9(c)(3), 1505.1, 1506.6(e), and 1508.4.

(2) Specific criteria for and identification of those typical classes of action:

(i) Which normally do require environmental impact statements.

(ii) Which normally do not require either an environmental impact statement or an environmental assessment (categorical exclusions (§ 1508.4)).

(iii) Which normally require environmental assessments but not necessarily environmental impact statements.

(c) Agency procedures may include specific criteria for providing limited exceptions to the provisions of these regulations for classified proposals. They are proposed actions which are specifically authorized under criteria established by an Executive Order or statute to be kept secret in the interest of national defense or foreign policy and are in fact properly classified pursuant to such Executive Order or statute. Environmental assessments and environmental impact statements which address classified proposals may be safeguarded and restricted from public dissemination in accordance with agencies' own regulations applicable to classified information. These documents may be organized so that classified portions can be included as annexes, in order that the unclassified portions can be made available to the public.

(d) Agency procedures may provide for periods of time other than those presented in § 1506.10 when necessary to comply with other specific statutory requirements.

(e) Agency procedures may provide that where there is a lengthy period between the agency's decision to prepare an environmental impact statement and the time of actual preparation, the notice of intent required by § 1501.7 may be published at a reasonable time in advance of preparation of the draft statement.

PART 1508—TERMINOLOGY AND INDEX

Sec.
1508.1 Terminology.
1508.2 Act.
1508.3 Affecting.
1508.4 Categorical Exclusion.
1508.5 Cooperating Agency.
1508.6 Council.
1508.7 Cumulative Impact.
1508.8 Effects.
1508.9 Environmental Assessment.
1508.10 Environmental Document.
1508.11 Environmental Impact Statement.
1508.12 Federal Agency.
1508.13 Finding of No Significant Impact.
1508.14 Human Environment.
1508.15 Jurisdiction By Law.
1508.16 Lead Agency.
1508.17 Legislation.
1508.18 Major Federal Action.
1508.19 Matter.
1508.20 Mitigation.
1508.21 NEPA Process.
1508.22 Notice of Intent.
1508.23 Proposal.
1508.24 Referring Agency.
1508.25 Scope.
1508.26 Special Expertise.
1508.27 Significantly.
1508.28 Tiering.

AUTHORITY: NEPA, the Environmental Quality Improvement Act of 1970, as amended (42 U.S.C. 4371 et seq.), Section 309 of the Clean Air Act, as amended (42 U.S.C. 7609), and Executive Order 11514, Protection and Enhancement of Environmental Quality (March 5, 1970, as amended by Executive Order 11991, May 24, 1977).

§ 1508.1 Terminology.

The terminology of this part shall be uniform throughout the Federal Government.

§ 1508.2 Act.

"Act" means the National Environmental Policy Act, as amended (42 U.S.C. 4321, et seq.) which is also referred to as "NEPA."

§ 1508.3 Affecting.

"Affecting" means will or may have an effect on.

§ 1508.4 Categorical exclusion.

"Categorical Exclusion" means a category of actions which do not individually or cumulatively have a significant effect on the human environment and which have been found to have no such effect in procedures adopted by a Federal agency in implementation of these regulations (§ 1507.3) and for which, therefore, neither an environmental assessment nor an environmental impact statement is required. An agency may decide in its procedures or otherwise, to prepare environmental assessments for the reasons stated in § 1508.9 even though it is not required to do so. Any procedures under this section shall provide for extraordinary circumstances in which a normally excluded action may have a significant environmental effect.

§ 1508.5 Cooperating agency.

"Cooperating Agency" means any Federal agency other than a lead agency which has jurisdiction by law or special expertise with respect to any environmental impact involved in a proposal (or a reasonable alternative) for legislation or other major Federal action significantly affecting the quality of the human environment. The selection and responsibilities of a cooperating agency are described in § 1501.6. A State or local agency of similar qualifications or, when the effects are on a reservation, an Indian Tribe, may by agreement with the lead agency become a cooperating agency.

§ 1508.6 Council.

"Council" means the Council on Environmental Quality established by Title II of the Act.

§ 1508.7 Cumulative impact.

"Cumulative impact" is the impact on the environment which results from the incremental impact of the action when added to other past, present, and reasonably foreseeable future actions regardless of what agency (Federal or non-Federal) or person undertakes such other actions. Cumulative impacts can result from individually minor but collectively significant actions taking place over a period of time.

§ 1508.8 Effects.

"Effects" include:

(a) Direct effects, which are caused by the action and occur at the same time and place.

(b) Indirect effects, which are caused by the action and are later in time or farther removed in distance, but are still reasonably foreseeable. Indirect effects may include growth inducing effects and other effects related to induced changes in the pattern of land use, population density or growth rate, and related effects on air and water and other natural systems, including ecosystems.

Effects and impacts as used in these regulations are synonymous. Effects includes ecological (such as the effects on natural resources and on the components, structures, and functioning of affected ecosystems), aesthetic, historic, cultural, economic, social, or health, whether direct, indirect, or cumulative. Effects may also include those resulting from actions which may have both beneficial and detrimental effects, even if on balance the agency believes that the effect will be beneficial.

§ 1508.9 Environmental assessment.

"Environmental Assessment":

(a) Means a concise public document for which a Federal agency is responsible that serves to:

(1) Briefly provide sufficient evidence and analysis for determining whether to prepare an environmental impact statement or a finding of no significant impact.

(2) Aid an agency's compliance with the Act when no environmental impact statement is necessary.

(3) Facilitate preparation of a statement when one is necessary.

(b) Shall include brief discussions of the need for the proposal, of alternatives as required by sec. 102(2)(E), of the environmental impacts of the proposed action and alternatives, and a listing of agencies and persons consulted.

§ 1508.10 Environmental document.

"Environmental document" includes the documents specified in § 1508.9 (environmental assessment), § 1508.11 (environmental impact statement), § 1508.13 (finding of no significant impact), and § 1508.22 (notice of intent).

§ 1508.11 Environmental impact statement.

"Environmental Impact Statement" means a detailed written statement as required by Sec. 102(2)(C) of the Act.

§ 1508.12 Federal agency.

"Federal agency" means all agencies of the Federal Government. It does not mean the Congress, the Judiciary, or the President, including the performance of staff functions for the President in his Executive Office. It also includes for purposes of these regulations States and units of general local government and Indian tribes assuming NEPA responsibilities under section 104(h) of the Housing and Community Development Act of 1974.

§ 1508.13 Finding of no significant impact.

"Finding of No Significant Impact" means a document by a Federal agency briefly presenting the reasons why an action, not otherwise excluded (§ 1508.4), will not have a significant effect on the human environment and for which an environmental impact statement therefore will not be prepared. It shall include the environmental assessment or a summary of it and shall note any other environmental documents related to it (§ 1501.7(a)(5)). If the assessment is included, the finding need not repeat any of the discussion in the assessment but may incorporate it by reference.

§ 1508.14 Human Environment.

"Human Environment" shall be interpreted comprehensively to include the natural and physical environment and the relationship of people with that environment. (See the definition of "effects" (§ 1508.8).) This means that economic or social effects are not intended by themselves to require preparation of an environmental impact statement. When an environmental impact statement is prepared and economic or social and natural or physical environmental effects are interrelated, then the environmental impact statement will discuss all of these effects on the human environment.

§ 1508.15 Jurisdiction By Law.

"Jurisdiction by law" means agency authority to approve, veto, or finance all or part of the proposal.

§ 1508.16 Lead agency.

"Lead Agency" means the agency or agencies preparing or having taken primary responsibility for preparing the environmental impact statement.

§ 1508.17 Legislation.

"Legislation" includes a bill or legislative proposal to Congress developed by or with the significant cooperation and support of a Federal agency, but does not include requests for appropriations. The test for significant cooperation is whether the proposal is in fact predominantly that of the agency rather than another source. Drafting does not by itself constitute significant cooperation. Proposals for legislation include requests for ratification of treaties. Only the agency which has primary responsibility for the subject matter involved will prepare a legislative environmental impact statement.

§ 1508.18 Major Federal action.

"Major Federal action" includes actions with effects that may be major and which are potentially subject to Federal control and responsibility. Major reinforces but does not have a meaning independent of significantly (§ 1508.27). Actions include the circumstance where the responsible officials fail to act and that failure to act is reviewable by courts or administrative tribunals under the Administrative Procedure Act or other applicable law as agency action.

(a) Actions include new and continuing activities, including projects and programs entirely or partly financed, assisted, conducted, regulated, or approved by federal agencies; new or revised agency rules, regulations, plans, policies, or procedures; and legislative proposals (§§ 1506.8, 1508.17). Actions do not include funding assistance solely in the form of general revenue sharing funds, distributed under the State and Local Fiscal Assistance Act of 1972, 31 U.S.C. 1221 et seq., with no Federal agency control over the subsequent use of such funds. Actions do not include bringing judicial or administrative civil or criminal enforcement actions.

(b) Federal actions tend to fall within one of the following categories:

(1) Adoption of official policy, such as rules, regulations, and interpretations adopted pursuant to the Administrative Procedure Act, 5 U.S.C. 551 et seq.; treaties and international conventions or agreements; formal documents establishing an agency's policies which will result in or substantially alter agency programs.

(2) Adoption of formal plans, such as official documents prepared or approved by federal agencies which guide or prescribe alternative uses of federal resources, upon which future agency actions will be based.

(3) Adoption of programs, such as a group of concerted actions to implement a specific policy or plan; systematic and connected agency decisions allocating agency resources to implement a specific statutory program or executive directive.

(4) Approval of specific projects, such as construction or management activities located in a defined geographic area. Projects include actions approved by permit or other regulatory decision as well as federal and federally assisted activities.

§ 1508.19 Matter.

"Matter" includes for purposes of Part 1504:

(a) With respect to the Environmental Protection Agency, any proposed legislation, project, action or regulation as those terms are used in Section 309(a) of the Clean Air Act (42 U.S.C. 7609).

(b) With respect to all other agencies, any proposed major federal action to which section 102(2)(C) of NEPA applies.

§ 1508.20 Mitigation.

"Mitigation" includes:

(a) Avoiding the impact altogether by not taking a certain action or parts of an action.

(b) Minimizing impacts by limiting the degree or magnitude of the action and its implementation.

(c) Rectifying the impact by repairing, rehabilitating, or restoring the affected environment.

(d) Reducing or eliminating the impact over time by preservation and maintenance operations during the life of the action.

(e) Compensating for the impact by replacing or providing substitute resources or environments.

§ 1508.21 NEPA process.

"NEPA process" means all measures necessary for compliance with the requirements of Section 2 and Title I of NEPA.

§ 1508.22 Notice of intent.

"Notice of Intent" means a notice that an environmental impact statement will be prepared and considered. The notice shall briefly:

(a) Describe the proposed action and possible alternatives.

(b) Describe the agency's proposed scoping process including whether, when, and where any scoping meeting will be held.

(c) State the name and address of a person within the agency who can answer questions about the proposed action and the environmental impact statement.

§ 1508.23 Proposal.

"Proposal" exists at that stage in the development of an action when an agency subject to the Act has a goal and is actively preparing to make a decision on one or more alternative means of accomplishing that goal and the effects can be meaningfully evaluated. Preparation of an environmental impact statement on a proposal should be timed (§ 1502.5) so that the final statement may be completed in time for the statement to be included in any recommendation or report on the proposal. A proposal may exist in fact as well as by agency declaration that one exists.

§ 1508.24 Referring agency.

"Referring agency" means the federal agency which has referred any matter to the Council after a determination that the matter is unsatisfactory from the standpoint of public health or welfare or environmental quality.

§ 1508.25 Scope.

Scope consists of the range of actions, alternatives, and impacts to be considered in an environmental impact statement. The scope of an individual statement may depend on its relationships to other statements (§§1502.20 and 1508.28). To determine the scope of environmental impact statements, agencies shall consider 3 types of actions, 3 types of alternatives, and 3 types of impacts. They include:

(a) Actions (other than unconnected single actions) which may be:

(1) Connected actions, which means that they are closely related and therefore should be discussed in the same impact statement. Actions are connected if they:

(i) Automatically trigger other actions which may require environmental impact statements.

(ii) Cannot or will not proceed unless other actions are taken previously or simultaneously.

(iii) Are interdependent parts of a larger action and depend on the larger action for their justification.

(2) Cumulative actions, which when viewed with other proposed actions have cumulatively significant impacts and should therefore be discussed in the same impact statement.

(3) Similar actions, which when

viewed with other reasonably foreseeable or proposed agency actions, have similarities that provide a basis for evaluating their environmental consequencies together, such as common timing or geography. An agency may wish to analyze these actions in the same impact statement. It should do so when the best way to assess adequately the combined impacts of similar actions or reasonable alternatives to such actions is to treat them in a single impact statement.

(b) Alternatives, which include: (1) No action alternative. (2) Other reasonable courses of actions. (3) Mitigation measures (not in the proposed action).

(c) Impacts, which may be: (1) Direct. (2) Indirect. (3) Cumulative.

§ 1508.26 Special expertise.

"Special expertise" means statutory responsibility, agency mission, or related program experience.

§ 1508.27 Significantly.

"Significantly" as used in NEPA requires considerations of both context and intensity:

(a) *Context.* This means that the significance of an action must be analyzed in several contexts such as society as a whole (human, national), the affected region, the affected interests, and the locality. Significance varies with the setting of the proposed action. For instance, in the case of a site-specific action, significance would usually depend upon the effects in the locale rather than in the world as a whole. Both short- and long-term effects are relevant.

(b) *Intensity.* This refers to the severity of impact. Responsible officials must bear in mind that more than one agency may make decisions about partial aspects of a major action. The following should be considered in evaluating intensity:

(1) Impacts that may be both beneficial and adverse. A significant effect may exist even if the Federal agency believes that on balance the effect will be beneficial.

(2) The degree to which the proposed action affects public health or safety.

(3) Unique characteristics of the geographic area such as proximity to historic or cultural resources, park lands, prime farmlands, wetlands, wild and scenic rivers, or ecologically critical areas.

(4) The degree to which the effects on the quality of the human environment are likely to be highly controversial.

(5) The degree to which the possible effects on the human environment are highly uncertain or involve unique or unknown risks.

(6) The degree to which the action may establish a precedent for future actions with significant effects or represents a decision in principle about a future consideration.

(7) Whether the action is related to other actions with individually insignificant but cumulatively significant impacts. Significance exists if it is reasonable to anticipate a cumulatively significant impact on the environment. Significance cannot be avoided by terming an action temporary or by breaking it down into small component parts.

(8) The degree to which the action may adversely affect districts, sites, highways, structures, or objects listed in or eligible for listing in the National Register of Historic Places or may cause loss or destruction of significant scientific, cultural, or historical resources.

(9) The degree to which the action may adversely affect an endangered or threatened species or its habitat that has been determined to be critical under the Endangered Species Act of 1973.

(10) Whether the action threatens a violation of Federal, State, or local law or requirements imposed for the protection of the environment.

§ 1508.28 Tiering.

"Tiering" refers to the coverage of general matters in broader environmental impact statements (such as national program or policy statements) with subsequent narrower statements or environmental analyses (such as regional or basinwide program statements or ultimately site-specific statements) incorporating by reference the general discussions and concentrating solely on the issues specific to the statement subsequently prepared. Tiering is appropriate when the sequence of statements or analyses is:

(a) From a program, plan, or policy environmental impact statement to a program, plan, or policy statement or analysis of lesser scope or to a site-specific statement or analysis.

(b) From an environmental impact statement on a specific action at an early stage (such as need and site selection) to a supplement (which is preferred) or a subsequent statement or analysis at a later stage (such as environmental mitigation). Tiering in such cases is appropriate when it helps the lead agency to focus on the issues which are ripe for decision and exclude from consideration issues already decided or not yet ripe.

Index

Act	1508.2
Action	1508.18, 1508.25
Action-forcing	1500.1, 1502.1
Adoption	1500.4(n), 1500.5(h), 1506.3
Affected Environment	1502.10(f), 1502.15
Affecting	1502.3, 1508.3
Agency Authority	1500.6
Agency Capability	1501.2(a), 1507.2
Agency Compliance	1507.1
Agency Procedures	1505.1, 1507.3
Agency Responsibility	1506.5
Alternatives	1501.2(c), 1502.2, 1502.10(e), 1502.14, 1505.1(e), 1505.2, 1507.2(d), 1508.25(b)
Appendices	1502.10(k), 1502.18, 1502.24
Applicant	1501.2(d)(1), 1501.4(b), 1501.8(a), 1502.19(b), 1503.1(a)(3), 1504.3(e), 1506.1(d), 1506.5(a), 1506.5(b)
Apply NEPA Early in the Process.	1501.2
Categorical Exclusion	1500.4(p), 1500.5(k), 1501.4(a), 1507.3(b), 1508.4
Circulating of Environmental Impact Statement.	1502.19, 1506.3
Classified Information	1507.3(c)
Clean Air Act	1504.1, 1508.19(a)
Combining Documents	1500.4(o), 1500.5(l), 1506.4
Commenting	1502.19, 1503.1, 1503.2, 1503.3, 1503.4, 1506.6(f)
Consultation Requirement.	1500.4(k), 1500.5(g), 1501.7(a)(6), 1502.25
Context	1508.27(a)
Cooperating Agency	1500.5(b), 1501.1(b), 1501.5(c), 1501.5(f), 1501.6, 1503.1(a)(1), 1503.2, 1503.3, 1506.3(c), 1506.5(a), 1508.5
Cost-Benefit	1502.23
Council on Environmental Quality.	1500.3, 1501.5(e), 1501.5(f), 1501.6(c), 1502.9(c)(4), 1504.1, 1504.2, 1504.3, 1506.6(f), 1506.9, 1506.10(e), 1506.11, 1507.3, 1508.6, 1508.24
Cover Sheet	1502.10(a), 1502.11
Cumulative Impact	1508.7, 1508.25(a), 1508.25(c)
Decisionmaking	1505.1, 1506.1
Decision points	1505.1(b)
Dependent	1508.25(a)
Draft Environmental Impact Statement.	1502.9(a)
Early Application of NEPA.	1501.2
Economic Effects	1508.8
Effective Date	1506.12
Effects	1502.16, 1508.8
Emergencies	1506.11
Endangered Species Act	1502.25, 1508.27(b)(9)
Energy	1502.16(e)
Environmental Assessment.	1501.3, 1501.4(b), 1501.4(c), 1501.7(b)(3), 1506.2(b)(4), 1506.5(b), 1508.4, 1508.9, 1508.10, 1508.13
Environmental Consequences.	1502.10(g), 1502.16
Environmental Consultation Requirements.	1500.4(k), 1500.5(g), 1501.7(a)(6), 1502.25, 1503.3(c)
Environmental Documents.	1508.10
Environmental Impact Statement.	1500.4, 1501.4(c), 1501.7, 1501.3, 1502.1, 1502.2, 1502.3, 1502.4, 1502.5, 1502.6, 1502.7, 1502.8, 1502.9, 1502.10, 1502.11, 1502.12, 1502.13, 1502.14, 1502.15, 1502.16, 1502.17, 1502.18, 1502.19, 1502.20, 1502.21, 1502.22, 1502.23, 1502.24, 1502.25, 1506.2(b)(4), 1506.3, 1506.8, 1508.11
Environmental Protection Agency.	1502.11(f), 1504.1, 1504.3, 1506.7(c), 1506.9, 1506.10, 1508.19(a)
Environmental Review Requirements.	1500.4(k), 1500.5(g), 1501.7(a)(6), 1502.25, 1503.3(c)
Expediter	1501.8(b)(2)
Federal Agency	1508.12
Filing	1506.9
Final Environmental Impact Statement.	1502.9(b), 1503.1, 1503.4(b)
Finding of No Significant Impact.	1500.3, 1500.4(q), 1500.5(1), 1501.4(e), 1508.13
Fish and Wildlife Coordination Act.	1502.25
Format for Environmental Impact Statement.	1502.10
Freedom of Information Act.	1506.6(f)
Further Guidance	1506.7
Generic	1502.4(c)(2)
General Services Administration.	1506.8(b)(5)
Geographic	1502.4(c)(1)
Graphics	1502.8
Handbook	1506.7(a)
Housing and Community Development Act.	1506.12, 1508.12
Human Environment	1502.3, 1502.22, 1508.14
Impacts	1508.8, 1508.25(c)
Implementing the Decision.	1505.3
Incomplete or Unavailable Information.	1502.22
Incorporation by Reference.	1500.4(j), 1502.21
Index	1502.10(j)
Indian Tribes	1501.2(d)(2), 1501.7(a)(1), 1502.15(c), 1503.1(a)(2)(ii), 1506.6(b)(3)(ii), 1508.5, 1508.12

Index

Intensity	1508.27(b)
Interdisciplinary Preparation.	1502.6, 1502.17
Interim Actions	1506.1
Joint Lead Agency	1501.5(b), 1506.2
Judicial Review	1500.3
Jurisdication by Law	1508.15
Lead Agency	1500.5(c), 1501.1(c), 1501.5, 1501.6, 1501.7, 1501.8, 1504.3, 1506.2(b)(4), 1506.8(a), 1506.10(e), 1508.16
Legislation	1500.5(j), 1502.3, 1506.8, 1508.17, 1508.18(a)
Limitation on Action During NEPA Process.	1506.1
List of Preparers	1502.10(h), 1502.17
Local or State	1500.4(n), 1500.5(h), 1501.2(d)(2), 1501.5(b), 1501.5(d), 1501.7(a)(1), 1501.8(c), 1502.16(c), 1503.1(a)(2), 1506.2(b), 1506.6(b)(3), 1508.5, 1508.12, 1508.18
Major Federal Action	1502.3, 1508.18
Mandate	1500.3
Matter	1504.1, 1504.2, 1504.3, 1508.19
Methodology	1502.24
Mitigation	1502.14(h), 1502.16(h), 1503.3(d), 1505.2(c), 1505.3, 1508.20
Monitoring	1505.2(c), 1505.3
National Historic Preservation Act.	1502.25
National Register of Historical Places.	1508.27(b)(8)
Natural or Depletable Resource Requirements.	1502.16(f)
Need for Action	1502.10(d), 1502.13
NEPA Process	1508.21
Non-Federal Sponsor	1501.2(d)
Notice of Intent	1501.7, 1507.3(e), 1508.22
OMB Circular A-95	1503.1(a)(2)(iii), 1505.2, 1506.6(b)(3)(i)
102 Monitor	1506.6(b)(2), 1506.7(c)
Ongoing Activities	1506.12
Page Limits	1500.4(a), 1501.7(b), 1502.7
Planning	1500.5(a), 1501.2(b), 1502.4(a), 1508.18
Policy	1500.2, 1502.4(b), 1508.18(a)
Program Environmental Impact Statement.	1500.4(i), 1502.4, 1502.20, 1508.18
Programs	1502.4, 1508.18(b)
Projects	1508.18
Proposal	1502.4, 1502.5, 1506.8, 1508.23
Proposed Action	1502.10(e), 1502.14, 1506.2(c)

Index

Public Health and Welfare.	1504.1
Public Involvement	1501.4(e), 1503.1(a)(3), 1506.6
Purpose	1500.1, 1501.1, 1502.1, 1504.1
Purpose of Action	1502.10(d), 1502.13
Record of Decision	1505.2, 1506.1
Referrals	1504.1, 1504.2, 1504.3, 1506.3(d)
Referring Agency	1504.1, 1504.2, 1504.3
Response to Comments	1503.4
Rural Electrification Administration.	1506.1(d)
Scientific Accuracy	1502.24
Scope	1502.4(a), 1502.9(a), 1508.25
Scoping	1500.4(b), 1501.1(d), 1501.4(d), 1501.7, 1502.9(a), 1506.8(a)
Significantly	1502.3, 1508.27
Similar	1508.25
Small Business Associations.	1506.6(b)(3)(vi)
Social Effects	1508.8
Special Expertise	1508.26
Specificity of Comments	1500.4(1), 1503.3
State and Areawide Clearinghouses.	1501.4(e)(2), 1503.1(a)(2)(iii), 1506.6(b)(3)(i)
State and Local	1500.4(n), 1500.5(h), 1501.2(d)(2), 1501.5(b), 1501.5(d), 1501.7(a)(1), 1501.8(c), 1502.16(c), 1503.1(a)(2), 1506.2(b), 1506.6(b)(3), 1508.5, 1508.12, 1508.18
State and Local Fiscal Assistance Act.	1508.18(a)
Summary	1500.4(h), 1502.10(b), 1502.12
Supplements to Environmental Impact Statements.	1502.9(c)
Table of Contents	1502.10(c)
Technological Development.	1502.4(c)(3)
Terminology	1508.1
Tiering	1500.4(i), 1502.4(d), 1502.20, 1508.28
Time Limits	1500.5(e), 1501.1(e), 1501.7(b)(2), 1501.8
Timing	1502.4, 1502.5, 1506.10
Treaties	1508.17
When to Prepare an Environmental Impact Statement.	1501.3
Wild and Scenic Rivers Act.	1506.8(b)(ii)
Wilderness Act	1506.8(b)(ii)
Writing	1502.8

APPENDIX D: The Clean Air Act §309.

§ 7609. Policy review

(a) The Administrator shall review and comment in writing on the environmental impact of any matter relating to duties and responsibilities granted pursuant to this chapter or other provisions of the authority of the Administrator, contained in any (1) legislation proposed by any Federal department or agency, (2) newly authorized Federal projects for construction and any major Federal agency action (other than a project for construction) to which section 4332(2)(C) of this title applies, and (3) proposed regulations published by any department or agency of the Federal Government. Such written comment shall be made public at the conclusion of any such review.

(b) In the event the Administrator determines that any such legislation, action, or regulation is unsatisfactory from the standpoint of public health or welfare or environmental quality, he shall publish his determination and the matter shall be referred to the Council on Environmental Quality.

July 14, 1955, c. 360, § 309, as added Dec. 31, 1970, Pub. L. 91-604 § 12(a), 42 U.S.C. § 7609 (1970).

APPENDIX E: Executive Order 11514 (as amended by Executive Order 11911) and Executive Order 12114

Executive Order 11514. March 5, 1970
PROTECTION AND ENHANCEMENT OF ENVIRONMENTAL QUALITY
As amended by Executive Order 11991. (Secs. 2(g) and (3(h)). May 24, 1977*

By virtue of the authority vested in me as President of the United States and in furtherance of the purpose and policy of the National Environmental Policy Act of 1969 (Public Law No. 91-190, approved January 1, 1970), it is ordered as follows:

Section 1. *Policy.* The Federal Government shall provide leadership in protecting and enhancing the quality of the Nation's environment to sustain and enrich human life. Federal agencies shall initiate measures needed to direct their policies, plans and programs so as to meet national environmental goals. The Council on Environmental Quality, through the Chairman, shall advise and assist the President in leading this national effort.

Sec. 2. *Responsibilities of Federal agencies.* Consonant with Title I of the National Environmental Policy Act of 1969, hereafter referred to as the "Act", the heads of Federal agencies shall:

(a) Monitor, evaluate, and control on a continuing basis their agencies' activities so as to protect and enhance the quality of the environment. Such activities shall include those directed to controlling pollution and enhancing the environment and those designed to accomplish other program objectives which may affect the quality of the environment. Agencies shall develop programs and measures to protect and enhance environmental quality and shall assess progress in meeting the specific objectives of such activities. Heads of agencies shall consult with appropriate Federal, State and local agencies in carrying out their activities as they affect the quality of the environment.

(b) Develop procedures to ensure the fullest practicable provision of timely public information and understanding of Federal plans and programs with environmental impact in order to

---

*The Preamble to Executive Order 11991 is as follows:

By virtue of the authority vested in me by the Constitution and statutes of the United States of America, and as President of the United States of America, in furtherance of the purpose and policy of the National Environmental Policy Act of 1969, as amended (42 U.S.C. 4321 *et seq.*), the Environmental Quality Improvement Act of 1970 (42 U.S.C. 4371 *et seq.*), and Section 309 of the Clean Air Act, as amended (42 U.S.C. 1857h-7), it is hereby ordered as follows:

obtain the views of interested parties. These procedures shall include, whenever appropriate, provision for public hearings, and shall provide the public with relevant information, including information on alternative courses of action. Federal agencies shall also encourage State and local agencies to adopt similar procedures for informing the public concerning their activities affecting the quality of the environment.

(c) Insure that information regarding existing or potential environmental problems and control methods developed as part of research, development, demonstration, test, or evaluation activities is made available to Federal agencies, States, counties, municipalities, institutions, and other entities, as appropriate.

(d) Review their agencies' statutory authority, administrative regulations, policies, and procedures, including those relating to loans, grants, contracts, leases, licenses, or permits, in order to identify any deficiencies or inconsistencies therein which prohibit or limit full compliance with the purposes and provisions of the Act. A report on this review and the corrective actions taken or planned, including such measures to be proposed to the President as may be necessary to bring their authority and policies into conformance with the intent, purposes, and procedures of the Act, shall be provided to the Council on Environmental Quality not later than September 1, 1970.

(e) Engage in exchange of data and research results, and cooperate with agencies of other governments to foster the purposes of the Act.

(f) Proceed, in coordination with other agencies, with actions required by section 102 of the Act.

(g) In carrying out their responsibilites under the Act and this Order, comply with the regulations issued by the Council except where such compliance would be inconsistent with statutory requirements.

Sec. 3. *Responsibilities of Council on Environmental Quality.* The Council on Environmental Quality shall:

(a) Evaluate existing and proposed policies and activities of the Federal Government directed to the control of pollution and the enhancement of the environment and to the accomplishment of other objectives which affect the quality of the environment. This shall include continuing review of procedures employed in the development and enforcement of Federal standards affecting environmental quality. Based upon such evaluations the Council shall, where appropriate, recommend to the President policies and programs to achieve more effective protection and enhancement of environmental quality and shall, where appropriate, seek resolution of significant environmental issues.

(b) Recommend to the President and to the agencies priorities among programs designed for the control of pollution and for enhancement of the environment.

(c) Determine the need for new policies and programs for dealing with environmental problems not being adequately addressed.

(d) Conduct, as it determines to be appropriate, public hearings or conferences on issues of environmental significance.

(e) Promote the development and use of indices and monitoring systems (1) to assess environmental conditions and trends, (2) to predict the environmental impact of proposed public and private

Section 309 of the Clean Air Act, as amended, for the Council's recommendation as to their prompt resolution.

(i) Issue such other instructions to agencies, and request such reports and other information from them, as may be required to carry out the Council's responsibilities under the Act.

(j) Assist the President in preparing the annual Environmental Quality Report provided for in section 201 of the Act.

(k) Foster investigations, studies, surveys, research, and analyses relating to (i) ecological systems and environmental quality, (ii) the impact of new and changing technologies thereon, and (iii) means of preventing or reducing adverse effects from such technologies.

Sec. 4. *Amendments of E.O. 11472.* Executive Order No. 11472 of May 29, 1969, including the heading thereof, is hereby amended:

(1) By substituting for the term "the Environmental Quality Council", wherever it occurs, the following: "the Cabinet Committee on the Environment".

(2) By substituting for the term "the Council", wherever it occurs, the following: "the Cabinet Committee".

(3) By inserting in subsection (f) of section 101, after "Budget,", the following: "the Director of the Office of Science and Technology,".

(4) By substituting for subsection (g) of section 101 the following:

"(g) The Chairman of the Council on Environmental Quality (established by Public Law 91-190) shall assist the President in directing the affairs of the Cabinet Committee."

(5) by deleting subsection (c) of section 102.

(6) By substituting for "the Office of Science and Technology", in section 104, the following: "the Council on Environmental Quality (established by Public Law 91-190)".

(7) By substituting for "(hereinafter referred to as the 'Committee')", in section 201, the following: "(hereinafter referred to as the 'Citizens' Committee')".

(8) By substituting for the term "the Committee", wherever it occurs, the following: "the Citizens' Committee".

actions, and (3) to determine the effectiveness of programs for protecting and enhancing environmental quality.

(f) Coordinate Federal programs related to environmental quality.

(g) Advise and assist the President and the agencies in achieving international cooperation for dealing with environmental problems, under the foreign policy guidance of the Secretary of State.

(h) Issue regulations to Federal agencies for the implementation of the procedural provisions of the Act (42 U.S.C. 4332(2)). Such regulations shall be developed after consultation with affected agencies and after such public hearings as may be appropriate. They will be designed to make the environmental impact statement process more useful to decisionmakers and the public; and to reduce paperwork and the accumulation of extraneous background data, in order to emphasize the need to focus on real environmental issues and alternatives. They will require impact statements to be concise, clear, and to the point,

Executive Order 12114. January 4, 1979
ENVIRONMENTAL EFFECTS ABROAD OF MAJOR FEDERAL ACTIONS

By virtue of the authority vested in me by the Constitution and the laws of the United States, and as President of the United States, in order to further environmental objectives consistent with the foreign policy and national security policy of the United States, it is ordered as follows:

Section 1.

1-1. Purpose and Scope. The purpose of this Executive Order is to enable responsible officials of Federal agencies having ultimate responsibility for authorizing and approving actions encompassed by this Order to be informed of pertinent environmental considerations and to take such considerations into account, with other pertinent considerations of national policy, in making decisions regarding such actions. While based on independent authority, this Order furthers the purpose of the National Environmental Policy Act and the Marine Protection Research and Sanctuaries Act and the Deepwater Port Act consistent with the foreign policy and national security policy of the United States, and represents the United States government's exclusive and complete determination of the procedural and other actions to be taken by Federal agencies to further the purpose of the National Environmental Policy Act, with respect to the environment outside the United States, its territories and possessions.

Sec. 2.

2-1. Agency Procedures. Every Federal agency taking major Federal actions encompassed hereby and not exempted herefrom having significant effects on the environment outside the geographical borders of the United States and its territories and possessions shall within eight months after the effective date of this Order have in effect procedures to implement this Order. Agencies shall consult with the Department of State and the Council on Environmental Quality concerning such procedures prior to placing them in effect.

2-2. Information Exchange. To assist in effectuating the foregoing purpose, the Department of State and the Council on Environmental Quality in collaboration with other interested Federal agencies and other nations shall conduct a program for exchange on a continuing basis of information concerning the environment. The objectives of this program shall be to provide information for use by decisionmakers, to heighten awareness of and interest in environmental concerns and, as appropriate, to facilitate environmental cooperation with foreign nations.

2-3. <u>Actions Included</u>. Agencies in their procedures under Section 2-1 shall establish procedures by which their officers having ultimate responsibility for authorizing and approving actions in one of the following categories encompassed by this Order, take into consideration in making decisions concerning such actions, a document described in Section 2-4(a):

(a) major Federal actions significantly affecting the environment of the global commons outside the jurisdiction of any nation (e.g., the oceans or Antarctica);

(b) major Federal actions significantly affecting the environment of a foreign nation not participating with the United States and not otherwise involved in the action;

(c) major Federal actions significantly affecting the environment of a foreign nation which provide to that nation:

- (1) a product, or physical project producing a principal product or an emission or effluent, which is prohibited or strictly regulated by Federal law in the United States becase its toxic effects on the environment create a serious public health risk; or

- (2) a physical project which in the United States is prohibited or strictly regulated by Federal law to protect the environment against radioactive substances.

(d) major Federal actions outside the United States, its territories and possessions which significantly affect natural or ecological resources of global importance designated for protection under this subsection by the President, or, in the case of such a resource protected by international agreement binding on the United States, by the Secretary of State. Recommendations to the President under this subsection shall be accompanied by the views of the Council on Environmental Quality and the Secretary of State.

2-4. <u>Applicable Procedures.</u> (a) There are the following types of documents to be used in connection with actions described in Section 2-3:

(i) environmental impact statements (including generic, program and specific statements);

(ii) bilateral or multilateral environmental studies, relevant or related to the proposed action, by the United States and one or more foreign nations, or by an international body or organization in which the United States is a member or participant; or

(iii) concise reviews of the environmental issues involved, including environmental assessments, summary environmental analyses or other appropriate documents.

(b) Agencies shall in their procedures provide for preparation of documents described in Section 2-4(a), with respect to actions described in Section 2-3, as follows:

(i) for effects described in Section 2-3(a), an environmental impact statement described in Section 2-4(a)(i);

(ii) for effects described in Section 2-3(b), a document described in Section 2-4(a)(ii) or (iii), as determined by the agency;

(iii) for effects described in Section 2-3(c), a document described in Section 2-4(a)(ii) or (iii), as determined by the agency;

(iv) for effects described in Section 2-3(d), a document described in Section 2-4(a)(i), (ii) or (iii), as determined by the agency.

Such procedures may provide that an agency need not prepare a new document when a document described in Section 2-4(a) already exists.

(c) Nothing in this Order shall serve to invalidate any existing regulations of any agency which have been adopted pursuant to court order or pursuant to judicial settlement of any case or to prevent any agency from providing in its procedures for measures in addition to those provided for herein to further the purpose of the National Environmental Policy Act and other environmental laws, including the Marine Protection Research and Sanctuaries Act and the Deepwater Port Act, consistent with the foreign and national security policies of the United States.

(d) Except as provided in Section 2-5(b), agencies taking action encompassed by this Order shall, as soon as feasible, inform other Federal agencies with relevant expertise of the availability of environmental documents prepared under this Order.

Agencies in their procedures under Section 2-1 shall make appropriate provision for determining when an affected nation shall be informed in accordance with Section 3-2 of this Order of the availability of environmental documents prepared pursuant to those procedures.

In order to avoid duplication of resources, agencies in their procedures shall provide for appropriate utilization of the resources of other Federal agencies with relevant environmental jurisdiction or expertise.

2-5. <u>Exemptions and Considerations.</u> (a) Notwithstanding Section 2-3, the following actions are exempt from this Order:

(i) actions not having a significant effect on the environment outside the United States as determined by the agency;

(ii) actions taken by the President;

(iii) actions taken by or pursuant to the direction of the President or Cabinet officer when the national security or interest is involved or when the action occurs in the course of an armed conflict;

(iv) intelligence activities and arms transfers;

(v) export licenses or permits or export approvals, and actions relating to nuclear activities except actions providing to a foreign nation a nuclear production or utilization facility as defined in the Atomic Energy Act of 1954, as amended, or a nuclear waste management facility;

(vi) votes and other actions in international conferences and organizations;

(vii) disaster and emergency relief action.

(b) Agency procedures under Section 2-1 implementing Section 2-4 may provide for appropriate modifications in the contents, timing and availability of documents to other affected Federal agencies and affected nations, where necessary to:

(i) enable the agency to decide and act promptly as and when required;

(ii) avoid adverse impacts on foreign relations or infringement in fact or appearance of other nations' sovereign responsibilities; or

(iii) ensure appropriate reflection of:

(1) diplomatic factors;

(2) international commercial, competitive and export promotion factors;

(3) needs for governmental or commercial confidentiality;

(4) national security considerations;

(5) difficulties of obtaining information and agency ability to analyze meaningfully environmental effects of a proposed action; and

(6) the degree to which the agency is involved in or able to affect a decision to be made.

(c) Agency procedures under Section 2-1 may provide for categorical exclusions and for such exemptions in addition to those specified in subsection (a) of this Section as may be necessary to meet emergency circumstances, situations involving exceptional foreign policy and national security sensitivities and other such special circumstances. In utilizing such additional exemptions agencies shall, as soon as feasible, consult with the Department of State and the Council on Environmental Quality.

(d) The provisions of Section 2-5 do not apply to actions described in Section 2-3(a) unless permitted by law.

Sec. 3.

3-1. Rights of Action. This Order is solely for the purpose of establishing internal procedures for Federal agencies to consider the significant effects of their actions on the environment outside the United States, its territories and possessions, and nothing in this Order shall be construed to create a cause of action.

3-2. Foreign Relations. The Department of State shall coordinate all communications by agencies with foreign governments concerning environmental agreements and other arrangements in implementation of this Order.

3-3. Multi-Agency Actions. Where more than one Federal agency is involved in an action or program, a lead agency, as determined by the agencies involved, shall have responsibility for implementation of this Order.

3-4. Certain Terms. For purposes of this Order, "environment" means the natural and physical environment and excludes social, economic and other environments; and an action significantly affects the environment if it does significant harm to the environment even though on balance the agency believes the action to be beneficial to the environment. The term "export approvals" in Section 2-5(a)(v) does not mean or include direct loans to finance exports.

3-5. Multiple Impacts. If a major Federal action having effects on the environment of the United States or the global commons requires preparation of an environmental impact statement, and if the action also has effects on the environment of a foreign nation, an environmental impact statement need not be prepared with respect to the effects on the environmental of the foreign nation.

JIMMY CARTER

THE WHITE HOUSE,
January 4, 1979

APPENDIX F: Relationships Between NEPA Requirements for EIS Contents and the Requirements of Procedures for Evaluation of Environmental Quality Objective

NEPA regulations requirements for EIS contents. (40 CFR 1502.10–1502.18)	Related requirements of these procedures. (18 CFR 714)
(a) *Cover sheet.* (40 CFR 1502.10(a) and 1502.11)	None.
(b) *Summary.* (40 CFR 1502.10(b) and 1502.12):	
(1) Major conclusions	714.432, Judge net EQ effects activity.
(2) Areas of controversy	714.411(c), Significance of EQ resources and attributes.
	714.433, Determine significant effects activity.
	714.441(c), Appraisal of effects on EQ attributes.
(3) Issues to be resolved	714.411(c), Significance of EQ resources and attributes.
	714.433, Determine significant effects activity.
	714.441(c), Appraisal of effects on EQ attributes.
(c) *Table of contents.* (40 CFR 1502.10(c))	None.
(d) *Purpose of and need for action.* (40 CFR 1502.10(d) and 1502.13).	None; but see P&S, 18 CFR 711.102.
(e) *Alternatives including proposed action.* (40 CFR 1502.10(e) and 1502.14):	
(1) Present effects in comparative form	None; but see P&S, 18 CFR 711, Subpart G.
(2) Explore and evaluate alternatives	Subpart C, General evaluation requirements:
	Subpart D, EQ evaluation process.
(3) Substantial treatment to each alternative considered in detail.	714.400(c)(1)(iii), Detailed definition-and-inventory stage.
	714.400(c)(1)(iv), Detailed assessment-and-appraisal stage.
(4) Include alternatives beyond agency jurisdiction	None; but see P&S, 18 CFR 711.50(c).
(5) Include no action	714.422, Forecast without-plans conditions activity.
(6) Identify preferred alternative(s)	None; but see P&S, 18 CFR 711.107.
(7) Include mitigation measures	None; but see P&S, 18 CFR 711.50(g).
(f) *Affected environment.* (40 CFR 1502.10(f) and 1502.15)	714.420, Inventory resources phase.
(g) *Environmental consequences.* (40 CFR 1502.10(g) and 1502.16):	
(1) Effects of alternatives	714.430, Assess effects phase.
	714.440, Appraise effects phase.
(2) Unavoidable adverse effects	714.440 Appraise effects phase.
(3) Relationship between local short-term uses of man's environment and maintenance and enhancement of long-term productivity.	714.432(b), Duration.
	714.432(c), Location.
(4) Irreversible and irretrievable commitments of resources	714.432(b), Duration.
(5) Direct effects	714.422, Forecast without-plans conditions activity.
	714.423, Forecast with-plan conditions activity.
(6) Indirect effects	714.422, Forecast without-plans conditions activity.
	714.423, Forecast with-plan conditions activity.
(7) Conflicts between the recommended plan (or candidate plans) and land use objectives.	714.441(c)(1)(iv), Institutional recognition.
(8) Energy requirements	None; but see P&S, 18 CFR 711.64(f).
(9) Natural or depletable resource requirements	Subpart D, EQ evaluation process.
(10) Urban quality, historic and cultural resources	Subpart D, EQ evaluation process.
(11) Mitigation means	None; but see P&S, 18 CFR 711.50(g).
(h) *List of preparers.* (40 CFR 1502.10(h) and 1502.17)	714.300, Interdisciplinary planning.
(i) *List of agencies, organizations, and individuals to whom copies of the statement are sent.* (40 CFR 1502.10(i)).	714.310, Public involvement.
(j) *Index.* (40 CFR 1502.10(j))	None.
(k) *Appendices.* (40 CFR 1502.10(k) and 1502.18)	714.330 Documentation.
	Appendix A, Example documentation formats.

Source: *Federal Register* 45(190):64446, 1980.

INDEX

Actions, 32, 40, 64
  characteristics, 40
  emergency, 31
  nonfederal, 24-25
  outside U.S., 17
  R&D, 19
  site-specific, 35
  stopping, 19-20
  temporary, 20-21
Administrative Procedures Act of 1946, 7, 12, 17-18
Air pollution, 9, 46, 49
Aliasing, 52
Alternatives, 16, 17, 19, 21, 22, 25, 38, 64, 65, 66-67
"Amchitka" (see *Committee for Nuclear Responsibility v. Seaborg*)
Ameliorative measures (see Mitigative measures and)
"Arbitrary and capricious," 18

Benefit-cost analysis, 5, 21, 23, 31
Biases, 23, 32-33, 38, 74
Biosphere (see Environment)
Bureau of Land Management, 4
Bureau of Reclamation, 4, 6

Calvert Cliffs Coordinating Committee v. Atomic Energy Commission, 19
Carter, President, 15, 64
Categorical exclusions, 20
Checklist(s), 36, 47
  Flow Chart and, 36-39
Clean Air Act of 1970, 16
Clean Water Act of 1972, 16
Coastal Zone Management Act of 1972, 16
Comments on EIS, 19, 21, 65, 75, 83
Committee for Nuclear Responsibility v. Seaborg, 19

Conservation Society of Southern Vermont v. The Secretary, 19
Constitution, U.S., 9, 29
Consultants, 51, 71, 73-75
Corruption, 76
Council on Environmental Quality, 5, 10, 12, 16, 17 (see also, Guidelines or Regulations)
Credentials, 69-70

Data Processing Service v. Camp, 18
DDT, 10
Development, 1, 19-20
Diversity, 23, 41, 45
Documentation, 22, 43, 62-63, 67

Earth Day, 3
Echo Park, 7
Ecosystem (see also Environment, biophysical), 9, 41, 45, 57, 58, 60
Endangered Species Act of 1976, 16
Enforcement, 11, 24-25, 29
Environmental Defense Fund v. Corps of Engineers, 19
Environment
  biosphere, 5, 40
  human, 18, 21
Environmental Protection Agency, 4, 9, 11, 16, 26, 51, 76
Environmental impact
  definition, 12, 44-45
  magnitude, 47
  prediciton of, 10, 43, 45
  probability of, 47, 50, 53
  significance of, 21, 39
  statements
    contents of, 55-56, 64-65
    definition, 21-22

form, 56-64
 style and format, 58-62
Environmental quality, 4, 20, 25-26, 39, 43
Erie, Lake, 10
Exemptions to EIS procedural requirements, 20-21
Exploitation, 3, 4

Federal Inter-Agency River Basin Committee, 6, 7-8
Federal Insecticide, Fungicide, and Rodenticide Act of 1948, 16
Federal Power Commission, 6, 19
Finding of No Significant Impact (see also Negative declaration), 21, 39
Fish and Wildlife Coordination Act of 1946, 7, 16
Food & Drug Administration, 4
Forest Service, 4, 76
Freedom of Information of 1967, 8, 18
Full disclosure, 11, 20, 21, 24, 29

Generalist, 66, 75
Geological Survey, 4
Gilham Dam (see *Environmental Defense Fund v. Corps of Engineers*)
Green Book, 5, 7
Greene County Planning Board v. Federal Power Commission, 19
Guidelines, CEQ, 12, 15, 17, 19, 64-65, 83
Gulf Oil (see *Natural Resources Defense Council v. Morton*)

Health and Human Services, Department of, 4
Health, environmental, 8, 11

Hearing, 21, 24, 27, 40, 71, 75, 82-84

Impact (see environment)
Industrial Revolution, 1
Inflation, 26
Interdisciplinary team and/or approach, 19, 23, 33, 40, 50, 53, 56
Intergovernmental Coordination Act of 1968, 8

Jackson, Senator, 8
Jargon, 38, 59, 70
Jefferson, Thomas, 9

Land use
 planning, 24
 spectrum, 41
Litigation of EISs, 18
"Little NEPA's", 23
Logic, formal, 60

Marine Protection, Research and Sanctuaries Act of 1976, 16
Matrices, 47-49, 66
Missouri River, 6
Mitigative (and ameliorative measures), 16, 21, 22, 39, 50, 51, 64, 69, 71
Muskie, Senator, 8

National Association of Environmental Professionals, 75
National Historic Preservation Act of 1976, 16
National Park Service, 4, 79
National Weather Service, 4
Navigable waters, 9
Negative declaration, 39, 56
NEPA
 as a comprehensive law, 4, 31

creating a process, 21
Natural Resources Defense Council v. Morton, 19
New York State
  Power Authority, 19
  State Environmental Quality Review Act of 1975, 16, 24, 64
Nixon, President, 7
Noise Control Act of 1976, 16
Null alternative, 21

Office of Environmental Quality, 12
Omnibus Flood Control Act of 1936, 5, 6, 7, 31
Omnibus Flood Control Act of 1944, 6
Oswego, New York, 53

Page limitation on EIS, 16, 63
Pennsylvania, 50
Pick-Sloan Plan, 6
Planning, 23, 32
Pollution (*see* specific type)
Preparers, list of, 16, 65
Preservation, 3, 18
President's Water Resources Council, 5, 8
Project Manager, 33, 35, 38, 75
Public involvement, 4, 18, 21, 23, 24, 25, 26, 29, 69, 70, 80, 81
Puerto Rico, 23

Reclamation Act of 1902, 6
Record of Decision, 16, 65
"Red Flag" review, 74
Refuse Act of 1899, 9
Regulations, CEQ, 12-17, 21, 22, 32, 39, 40, 56, 63, 64-65, 66, 71, 83
Reorganization Plans, 7
Responsible Federal Official, 19, 43, 71

Resource Conservation and Recovery Act of 1976, 16
Roosevelt, President F., 7

Safe Drinking Water Act of 1976, 16
Scenic Hudson Preservation Conference v. Federal Power Commission, 18
Scoping, 15-16
Senate Document No. 97, 5
Short-term uses v. long-term productivity, 3, 25-27, 38, 49, 64
"Silent Spring", 2, 10
Site visit, 35, 38, 70, 74
Soil Conservation Service, 4, 9
Soil pollution, 8
Stability, 45
Standing, 18, 19
Systems, environmental, 44-45

Television, 10
Tiering, 16, 60
Toxic Substances Control Act of 1976, 16

Urban problems and environmental quality, 2

Vietnam War, 10

Water pollution, 8-10, 46, 49, 53
Water Pollution Control Amendments of 1972, 16
Water Resources Council, 8, 17, 31, 49
Water Resources Planning Act of 1965, 5, 8
Witness stand, 70, 75
Worst Condition Analysis, 46, 51, 53

Zoning, 27, 35, 48

About the Author

PETER E. BLACK is Professor of Water and Related Land Resources at the State University of New York College of Environmental Science and Forestry in Syracuse, New York. He was a Research Forester with the U.S. Forest Service at the Coweeta Hydrologic Laboratory in North Carolina and taught at Humboldt State College in Arcata, California.

Dr. Black has published on a wide variety of topics related to water resources and environmental impact analysis, and has been project manager on numerous environmental assessment projects with Impact Consultants, of which he is a principal. He is active in the American Water Resources Association—of which he is a charter member—and he has served as chairman of its Education Committee, Coordinator of the Educational Water Film Festival, and District Director. He is a member of several other professional organizations, including the National Association of Environmental Professionals and the American Society of Professional Consultants. He is a Certified Environmental Professional by the NAEP.

Dr. Black received the B.S. and Master of Forestry degrees from the University of Michigan, and the PhD. degree from Colorado State University in Fort Collins, Colorado, in 1961. He is native of New York City.

THE LIBRARY
ST. MARY'S COLLEGE OF MARYLAND
ST. MARY'S CITY, MARYLAND 20686